ENERGY MANAGEMENT
Policy, Planning and Utilization

About the Author

Pradeep Chaturvedi, is a well known Expert in the Energy Conservation and Renewable Energy Sector. His expertise lies in project planning, implementation and management with emphasis on technology transfer and skill building. A number of pioneering projects and studies of national and international significance have been conducted under his guidance. He has planned, developed and implemented projects in the Asia-Pacific region.

Mr. Chaturvedi is the Vice Chairman of the World Renewable Energy Network supported by UNESCO and based in U.K. Being a founder of the Bio Energy Society of India, and now its President, he is involved in projecting the bio-fuels issue at national and international level. He is also Member of the Study Committee of the World Energy Council.

Mr. Chaturvedi is a Graduate in Engineering. He is a Fellow of the Institution of Engineers (India) and Fellow of the Institute of Directors. He is also the Recipient of FAO's World Food Day Medal 1991.

Mr. Chaturvedi has edited and authored 17 books related to sustainable development, renewable energy and energy conservation; and has contributed papers at a number of national and international conferences.

ENERGY MANAGEMENT
Policy, Planning and Utilization

PRADEEP CHATURVEDI
President
Bio-Energy Society of India

CONCEPT PUBLISHING COMPANY, NEW DELHI

ISBN 81-7022-669-4

First Published 1997

Published and Printed by
Ashok K. Mittal
CONCEPT PUBLISHING COMPANY
A/15-16, Commercial Block, Mohan Garden,
NEW DELHI - 110059 (India)

Phones: 5554042, 5504042
Fax : 91-11-5598898

Dedicated to
My Friend
IAN D. LINDSAY
Secretary General
World Energy Council

for

his outstanding contribution in building up the World Energy Council, his exemplary vision in bringing renewable energy sources centre stage; and focusing attention on the urgency for fulfilling energy needs of 2 billion poor in the developing countries-who do not have access to commercial energy.

Dedicated to

My Friend

IAN D. LINDSAY

Secretary General

World Energy Council

INTRODUCTION

Energy has been a primary concern of development planning. This sector has attracted upto 28 per cent of national Plan allocation, power sector getting upto 18 per cent. Increasing energy demands required additional huge investments, which was beyond government resources. To supplement its own funds the government promoted foreign and private participation, in 1992, initially in the power sector, and subsequently in renewable, petroleum and coal sectors. In March 1997, the government has also signed an agreement for installation of nuclear power plant with assistance from Russia. Capacity additions expected through such effort have not been at the desired level. The situation has also caused concern in the general user and upset schedule of the economic sectors—both industrial and agricultural. Supporters of privatisation blame lack of procedures to be the main cause for delays and consider privatisation as the ultimate saviour of the energy sector; and the supporters of old policy think that if the government had continued on the old path of capacity additions through plan funds, India could have been in a better position. These are the two extreme viewpoints, and the solution appears to lie somewhere in between.

A general consensus appears that the necessity of attaining higher economic growth rate during the Eighth Plan period prompted the government to undertake Fast Track Projects in the power generation sector. Lack of procedures and necessary infrastructure (both human and legal) to deal with the privatisation has caused embarrassment to the government. Gaining from the experience the government agencies are now in a better position to deal with foreign and private participation, which is coming in at a slow pace. Many of the issues have been identified for the situation and for some of them it appears that appropriate solutions have been found. However, a major concern for such power projects will be their dependence on imported fuel supply.

State Electricity Boards have been criticised for their poor financial situation. Critics often overlook that it is only in this sector that the input to farmers (i.e. electricity) is being subsidised by the supplier. All other agricultural inputs like seeds, fertilizer etc are subsidised by the Ministry of Agriculture and the producing industry gets full price. If the same norm is followed in the power sector and the tariff is subsidised by the Ministry of Agriculture, all the State Electricity Boards will receive their full cost of supply and their financial situation will change. Most of them will show their rate of return on investment higher than 3 per cent. That will place electricity boards in a better position to attract private and foreign investments.

The other sectors of primary fossil energy—coal and oil and natural gas are slowly opening up to privatisation. However, shortage of coal and oil has forced government to rely on import of low ash coal and import of increasing quantities of crude oil. Erratic power supply had resulted in customers using individual generators. It was estimated that during 1994-95 almost 1 mt oil was consumed in standby generators. Experts feel that this figure may rise to 3-4 million tonnes during 1997-98. Rising international market price of crude oil coupled with rupee devaluation has increased financial burden in terms of higher import bill. This situation is expected to aggravate further in the coming years. The present oil pool deficit of over Rs.15,000 crores, as on March 31, 1997, is already causing heavy stress on Indian economy.

Higher imports of fuels has raised concern about long term energy security. That brings in focus the issue of energy policy aimed at long term sustainable energy supply at an affordable cost. Immediate relief can be brought around only by intensifying energy conservation efforts, and higher exploitation of nuclear power and renewable energy sources, especially the biomass fuels. In these sectors, India has experience and strong base of technical manpower and know-how, and indigenous resource supply option.

Associated problems will get solutions only when large scale operations will be undertaken. Both these sectors have passed through a lean period, though for different reasons. These actors are now better geared for bigger job. Bio-Nuc scenario projected by the World Energy Council in its Study 'Global Energy Perspectives to 2050 and Beyond' provide an ideal framework for long term application. Closer interaction with the experts from WEC may offer options for energy policy.

Energy conservation efforts made by Energy Management Centre, EMC, and Petroleum Conservation Research Association, PCRA, through FICCI, CII and ASSOCHAM has resulted in considerable top management commitment of industry to energy conservation and consequent improvement in energy elasticities in the large scale industries. Twenty-five of the leading industries recording significant financial savings have received National Awards for Energy Conservation in the year 1996. The number of small and medium scale industries have also registered savings. Other sectors—transport, agriculture and domestic—have to now opt energy conservation measures on a large scale. Lack of awareness of conservation measures; local production of non-standardised and inefficient equipment; low energy supply price and considerably higher cost of energy conservation devices and equipment are a few identified reasons for slow movement of energy conservation measures. Such a movement can proceed at a rapid pace only when the energy pricing is done right and uninterrupted energy supply of high quality is assured. Till that time the effect of present efforts will be perceptible but not optimum. A combination of appropriate energy pricing and regulatory mechanism can certainly lead to energy savings of 10-15 per cent during the IX Five Year Plan (1997-2002).

India has agreed to purchase 3,000 MW equivalent power supply from Pakistan. This is a major breakthrough in sub-regional cooperation. ESCAP has suggested that sub-regional cooperation can help in improved energy supply. Nepal and Bhutan are identified as source of supply of hydro electricity to the east and north-east of India. Similarly, the natural gas pipeline from Oman to India can enhance gas supply possibility. Though such a suggestion makes long term economic sense but its implementation has many constraints, most important being the national security. This approach will provide an option for long term uninterrupted fuel supply.

Long term Energy Policy and flexibility in its implementation will be the prime consideration. Widest possible range of economic and publicly acceptable options have to be kept open—including accelerating renewable energy development and maintaining the nuclear option. As the lead times to achieve major changes are long, the action has to be taken now. Policies need clear time tables with achievable targets and action programmes. Appropriate actions will recognise current needs and realities—in terms of technology, finance and infrastructure.

Future energy development in India will have to consider the following elements:

i. Need to create/improve conditions for rural social and economic development through decentralised stand alone systems, primarily on renewable sources of energy. Current trend indicates that the population pressure will increase at a rapid pace in cities; and rural areas will experience lower population increase, and will be amenable to planned growth.

ii. Environmental considerations will dictate the energy option. Low pollution creating energy technologies will get precedence. Emphasis on local pollution will cover all conversion and end-use devices. Nuclear and bio fuels will find more support.

iii. Cost of end-use device and price of energy supply, both, will be important, thereby supporting end-use efficiency improvement.

iv. Promotion of foreign and private investment will be a thrust area, but not the only solution. Whereas efforts are to be made to motivate such investments, internal generation of funds by the public utilities will play an important role in power generation capacity additions. This will be through good house-keeping and appropriate energy pricing. The subsidy to electricity supplied to farmers should be provided by the Ministry of Agriculture; and the State Electricity Boards should be relieved of the burden.

v. Technological innovation and diffusion, together with supportive institutional changes, will have a powerful role to play in promoting sustainable energy development and use.

Action in the energy sector has to be planned on a long term basis—responding to the economic and social conditions. However, decisions in the energy sector will certainly be influenced by the above mentioned factors. A similar approach was also endorsed by the experts participating at the World Energy Council Forum on South Asia.

The book is a compilation of Papers submitted by the author at different conferences on energy, and provides reference information on the energy situation. As the papers were presented at different times they provide an overview of energy situation over a 5-7 years time frame. Various subjects like energy, research and development; total quality control in power sector; energy conservation; and energy and environment interrelation have been dealt carefully. Also subjects like regional and international cooperation are dealt in the current context. As use of biomass for energy and solar energy through thermal route are important for the country, the same have also been covered. Wherever possible case studies based on personal experience have also been added to indicate the situation under field conditions. Specific reference has been made to the experience gained in the Asia Pacific Region under the Commonwealth Project. Recommendations of the conferences, where the Papers were submitted, have also been annexed, (in selected cases), where it was considered that wider information on expert opinion on the topic will be relevant. A Chapter on Energy Situation in India has been compiled based on the recent reports to give an insight into the current situation.

Energy Data for India refers to the statistics prepared by the Planning Commission and relevant Ministries of the Government of India, and that for the other countries refer to the Resource Studies and Energy Data Profiles developed by the World Energy Council.

I hope the Book will serve as a reference volume to readers of different cross-sections of the society.

I have pleasure in dedicating this Book to my friend Mr. I.D.Lindsay, Secretary General, World Energy Council, WEC, for his outstanding contribution in building up the World Energy Council. He has shown exemplary vision in directing the WEC for a Decade—bringing renewable energy sources centre stage and focusing attention on the urgency for fulfilling energy needs of the 2 billion poor in the developing countries, who do not have access to commercial energy His contribution aimed at uplift of the humanity deserves to be inscribed in golden letters.

June 15, 1997 PRADEEP CHATURVEDI
New Delhi

ACKNOWLEDGEMENTS

I wish to gratefully acknowledge the help and inputs provided by Mr.J.N.Maggo, Joint Adviser, Planning Commission, Government of India.

I wish to thank Dr. Ranjit Watson, Chief Projects Officer, Industrial Development Unit, Commonwealth Secretariate, for all his help and support during the Commonwealth Energy Project work in the Asia- Pacific Region.

I am also grateful to Dr. Gustaro Best, Senior Energy Coordinator, FAO; and Prof. W.S. Hulscher, Chief Technical Adviser, FAO's Regional Wood Energy Development Programme in Asia for various inputs on bio-fuels; and to Mr. Michael Jefferson for elaboration on WEC Data Profiles.

Mr.Ashok Mittal, Managing Director, Concept Publishing Company deserves appreciation and thanks for undertaking publication of the Book.

June 15, 1997 PRADEEP CHATURVEDI

New Delhi

CONTENTS

1

ENERGY SITUATION IN INDIA

INTRODUCTION

Maintenance of adequate infrastructure facilities at affordable cost is crucial in ensuring rapid economic growth. Energy being an important element of the infrastructure sector has to be ensured on a sustainable basis. The government's development strategy is now shifting from public owned infrastructure creation to development of efficient infrastructure and towards creating an enabling environment for private participation. A variety of options are available for public-private partnership that may encourage better risk sharing, accountability, cost recovery and management of infrastructure. Thus, the government is now working more as a facilitator. Whereas this role relieves them of the responsibility of implementation, it makes imperative that the government takes careful decisions as thousands of small common stakeholders would be contributing in the process of development. The public investments will be more out of faith in government's policy of privatisation.

Also, the investments being of a long gestation period call for careful matching of government's policies and the expected returns. Whereas the fully government controlled infrastructure development was an administrative exercise, motivating private investments calls for financing from commercial sources that have many other options to invest their money. Proving bankability and soundness of the project is becoming all the more important. Given the competing demands on limited resources, the procedures relating to foreign investments in infrastructure are being liberalised.

ENERGY POLICY

A number of high level committees have been set up in the past by the Government of India to examine in detail the various issues and suggest measures for ensuring development of the energy sector for sustaining the economic growth and meeting the basic needs of the people. The recommendations made by these committees are kept in view while formulating the Five Year Plans. The 'strategy in energy sector' adopted for all Five Year Plans incorporates the changing needs of the economy and the society. It was felt desirable during the formulation of the 8th Five Year Plan (1992-97) to adopt a long-term strategy which is consistent with the goal of sustainable development.

The short-term and the long-term priorities as outlined in the Eighth Plan document are given as follows:

Short-Term

- maximise returns from the assets already created in the energy sector.
- initiate measures for reducing technical losses in production, transportation and end-use of all forms of energy.
- initiate action to reduce the energy intensity of the different consuming sectors of the economy and promote conservation and demand management through appropriate organisational and fiscal measures.
- initiate steps for meeting fully the basic energy needs of the rural and urban households, so as to reduce the existing inequities in energy use.
- maximise satisfaction of demand for energy from indigenous resources.

Medium-Term

- initiate steps towards progressive substitution of petroleum products by coal, lignite, natural gas and electricity so as to restrict the quantum of oil imports to the 1991 level.
- initiate action for accelerated development of all renewable energy resources, especially the available hydro-electric potential.
- promote programmes to achieve self-reliance in the energy sector.
- initiate appropriate organisational changes in the case of different energy sub-sectors consistent with the overall energy strategy.

Long-Term

— promote an energy supply system based largely on renewable sources of energy.

— promote technologies of production, transportation and end-use of energy that are environmentally benign and cost-effective.

These policy elements were built on the hypothesis that foreign and private investments would be forthcoming at desired levels. As lack of procedures to fulfil accountability norms have resulted in poor response from investors, the performance is not in conformity with the laid down policy.

PRIMARY ENERGY ENDOWMENTS

Present proven recoverable reserves of different sources of commercial energy in India (in 1993-94) are as follows:

TABLE 1.1: Recoverable Reserves of Commercial Energy in India

i.	Coal	68,047	mt
ii.	Crude Oil	757	mt
iii.	Natural Gas	686	bcum

Ref: Planning Commission, Government of India, 1997

Emerging technologies are likely to make higher amounts of recoverable reserves in the country.

Renewable Energy Potential

A resource survey has indicated large potential of the renewable energy sources. Following Table 1.2 gives an overall potential in the country in 1993-94:

The potential cannot be taken as exploitable target under existing technology know how. However, projection of this potential may lead to new technologies development in India.

PRIMARY ENERGY PRODUCTION AND CONSUMPTION PATTERN

Energy Production

Amongst the indigenously produced primary commercial fuels, the

TABLE 1.2: Renewable Energy Potential in India

Sources/Technology	Potential/ Availability
Biogas	12 million Nos.
Biomass	17,000 MW
High Efficiency Wood stove	120 million Nos.
Solar Energy	5 X 10 WHr/Year
Small Hydro	10,000 MW
Wind Energy	20,000 MW
Ocean Thermal	50,000 MW
Sea Wave Power	20,000 MW
Tidal Power	9,000 MW
Urban Municipal & Industrial Waste	1,000 MW

Ref: Planning Commission, Government of India, September, 1995.

relative share of oil and gas has increased from 1.2 per cent in 1950-51 to 32 per cent in 1990-91; whereas the share of coal has declined from 98 per cent in 1950-51 to 63 per cent in 1990-91. The share of coal is reflected to be 67 per cent in indigenously produced primary commercial fuels in 1995-96, mainly due to lower crude oil production.

Trends in indigenous production of commercial energy, for the period 1960-61 to 1995-96, are given in Table 1.3 below:

TABLE 1.3: Indigenous Commercial Energy Production

Source	Unit	1960-61	1970-71	1980-81	1990-91	1993-94 (*)	1995-96 (*)
Coal & Lignite	mt	55.72	76.34	118.81	225.50	264.14	292.27
Crude Oil & NGL	mt	0.45	6.82	10.51	33.02	27.76	35.20
Natural Gas	bcum	-	1.33	2.17	16.55	16.86	22.63
Hydro Electricity	MW(e)	1,900	6,400	11,800	18,800	20,400	21,000
Nuclear Power	MW(e)	-	400	900	1,500	2,250	2,250

Ref: Planning Commission, Government of India - September 1995. *Economic Survey of India, 1996-97.*

* Provisional Figures.

Energy Consumption

The average decennial rate of growth of commercial energy consumption during 1953-54 and 1990-91 varied between 4.21 per cent and 5.47 per cent. In view of the shortages and restrictions, the past trend of commercial energy consumption does not really represent the growth of demand for such energy but merely reflects the growth of its actual availability. Table 1.4 shows the average annual rate of growth in commercial energy consumption during the last four decades:

TABLE 1.4: Rate of Decennial Growth in Commercial Energy Consumption

Period	Rate of Growth (per cent per annum)
1953/54-1960/61	5.47
1960/61-1970/71	4.29
1970/71-1980/81	4.21
1980/81-1990/91	4.52

Ref: Eighth Plan Document - Planning Commission, Government of India, 1992.

The pattern of primary commercial energy production and consumption in the country has undergone a change over the last four decades as is evident from the energy balance given in Table 1.5. The total primary energy consumption (including commercial and non-commercial) in the economy has increased from 3,752 peta joules (PJ) in 1953 to 13,524 PJ in 1993. In terms of oil equivalent units, the total primary energy consumption has increased from 89.8 million tonnes of oil equivalent (mtoe) to 323.01 mtoe during the same period. The share of primary commercial energy consumption has gone up from 28.4 per cent to 63.2 per cent. The primary commercial energy consumption has grown at a rate of 5.34 per cent per annum as compared to a growth rate of 3.26 per cent in the total primary energy. As far as final energy consumption is concerned, it has increased from 3,537 PJ or 84.48 mtoe in 1953 to 10,569 PJ or 252.44 mtoe in 1993 at an annual compound rate of growth of 2.77 per cent. The rate of growth of final commercial energy consumption has been 4.81 per cent during this period. In absolute terms, the final commercial energy consumption in the economy has increased from 852 PJ (20.35 mtoe) in 1953 to 5,595 PJ (133.63 mtoe) in 1993. The household sector accounts for the entire final consumption of non-commercial energy.

The change in the relative share of commercial and non-commercial sources in the primary energy supplies and the share of different sources in the final commercial energy consumption are shown in Table 1.5.

TABLE 1.5: Change in Pattern of Energy Consumption (per cent)

	Primary Energy		Final Commercial Energy				
Year	Commer-cial	Non-com-mercial	Coal	Pet. Prod.	Nat. Gas	Elec-tricity	Total
1953	28.4	71.6	80.1	16.7	0.0	3.2	100.0
1960	34.7	65.3	75.3	19.9	0.0	4.8	100.0
1970	40.6	59.4	56.1	34.1	0.6	9.2	100.0
1980	47.5	52.5	47.9	40.3	1.1	10.7	100.0
1990	59.9	40.1	35.9	43.6	5.5	15.0	100.0
1993	63.2	36.8	32.0	44.5	6.4	17.1	100.0

Ref: J.N. Maggo - R&D Needs for Energy Sector - September 1996.

As may be seen from Table 1.6, the share of coal is declining in the final commercial energy consumption, and that of other fuels is increasing, particularly of oil products.

Table 1.6 gives the sourcewise consumption of final energy.

TABLE 1.6: Sourcewise Consumption of Final Energy (Peta Joules)

	Non-commercial Energy				Commercial Energy				
Year	Fuel-wood	Dung cake	Agro-Waste	Total	Coal & Lignite	Pet. Prod.	Nat. Gas	Elec. tricity	Total
1953	1700	466	519	2685	683	142	-	27	852
1960	1968	543	603	3114	952	252	-	61	1265
1970	2345	653	633	3631	1070	649	12	175	1906
1980	2864	794	774	4432	1444	1212	33	323	3011
1990	3210	890	874	4974	1827	2218	283	762	5090
1993	3210	890	874	4974	1792	2490	357	956	5595

Ref: J.N.Maggo - R&D Needs for Energy Sector - September, 1996.

There have also been changes over time in the sectoral use of different commercial fuels in the economy. These changes are shown in Table 1.7 which gives the percentage share of commercial fuel used in 1970 and 1993 in various sectors of consumption.

TABLE 1.7: Relative Shares of Commercial Fuels in Sectoral Consumption (per cent)

	Coal		Petro. Prod.		Natural Gas		Electricity	
Sector	1970	1993	1970	1993	1970	1993	1970	1993
Household	8.69	0.67	19.26	19.90	-	2.04	7.92	16.33
Commercial	-	-	2.31	0.74	-	-	5.31	5.33
Industry	60.47	91.57	26.19	9.71	41.67	20.64	70.75	45.72
Agriculture	-	-	5.39	10.42	-	-	9.22	26.63
Transport	30.47	2.68	31.74	44.26	-	-	2.88	2.12
Others	-	-	-	-	-	-	3.92	3.86
Feedstock	0.37	5.08	10.79	8.81	58.33	77.32	-	-
Non-Energy	-	-	4.31	6.16	-	-	-	-
TOTAL	100.0	100.0	100.0	100 .0	100.0	100.0	100.0	100.0

Ref: J.N. Maggo - R&D Needs for Energy Sector - September 1996.

The shares of different sectors in the consumption of commercial energy are given in Table 1.8.

TABLE 1.8: Share of Different Sectors in Consumption of Commercial Energy (per cent)

Sector	1953	1960	1970	1980	1990	1993
Household	9.6	9.7	12.2	9.7	11.8	12.0
Commercial	0.5	0.3	1.2	0.9	1.1	1.2
Industry	39.9	41.5	49.6	49.6	44.7	42.8
Agriculture	1.4	1.6	2.7	6.3	8.2	9.2
Transport	43.7	40.2	28.2	23.5	19.8	20.8
Others	1.5	1.7	0.4	0.4	0.9	0.9
Feedstock	0.5	0.3	4.2	5.6	10.4	10.4
Non-Energy	2.9	4.7	1.5	4.0	3.1	2.7

Ref: J.N. Maggo - R&D Needs for Energy Sector - September 1996.

Energy Intensity of the Economy

The change in the primary energy intensity of the economy over the years are given in Table 1.9.

TABLE 1.9: Change in overall Primary Energy Intensities w.r.t. GDP (PJ/US $ Billion)

	1953	1960	1970	1980	1990	1993
Non-Commercial	46.04	40.63	32.96	29.71	19.23	17.29
Commercial	18.30	21.55	22.57	26.89	28.75	29.73
Total	64.34	62.18	55.53	56.60	47.98	47.02

Ref: Planning Commission, Government of India - September 1996.

As may be seen from Table 1.9 above, the overall energy intensity of the economy has declined over the years. This has been made possible by the gradual substitution of primary non-commercial energy sources by the more efficient commercial energy sources. As a result the primary commercial energy intensity has gone up from 18.3 to 29.72 PJ/US $ billion during the period 1953 to 1993.

The coefficient of elasticity of final commercial energy consumption with respect to GDP at constant prices has been 1.14 during the period 1970-1993. With a substantial growth in the GDP in 1995-96, the energy elasticities in the industrial sector have improved substantially.

Details of elasticities for different decennial periods are given in Table 1.10.

TABLE 1.10 : Elasticities of Consumption of Commercial Energy w.r.t. GDP

Period	Elasticity of Consumption with reference to GDP			
	Coal	Oil	Electricity	Total Comm. Energy
1953/54-1960/61	1.10	2.14	3.02	1.37
1960/61-1970/71	0.53	2.31	3.04	1.16
1970/71-1980/81	0.98	1.83	2.06	1.37
1980/81-1990/91	0.31	1.12	1.57	0.82

Ref: Eighth Plan Document, Planning Commission, Government of India - 1992.

Non-Commercial Sources of Energy (Biomass only)

Traditionally the biomass based renewable energy sources contribute only to rural domestic energy supply. Other renewable energy sources contribute marginally in the overall energy scene. Therefore, only the biomass resources has been considered. Following Table 1.11 gives details of contribution of different biomass energy sources.

TABLE 1.11: Biomass Use as Energy Fuel in India (1991-92) (mt/year)

S.No.	Source	Availability	Estimated consumption	Deficit
I.	Fuelwood	28.4	166.85	138.40
II.	Urban/Indl.Waste	246.0	Negligible	—
III.	Agro Residue	240.0	150.0	—
IV.	Cattle Dung	960.0	80.0 (Dry Wt)	—

Ref: Planning Commission, Government of India - September, 1995.

FUTURE ENERGY DEMAND

Expert consultations at the Asia Energy Vision 2020, organised under the auspices of the World Energy Council, agreed on energy demand projection in India, upto 2020, as given in Table 1.12:

TABLE 1.12: Future Demand for Commercial Energy in India

Source	Unit	1991-92	1996-97	2009-10	2020-21*
1. Electricity	TWH	231	336	725	1300 (1745) **
2. Coal	mt	229	311	690@	1345
3. Petroleum Products	mt	57	81.2	165	335
4. Natural Gas	bcum	18.6	30.2	65	130

Ref: Planning Commission, Government of India -September 1995 & September 1996.

* Projections to 2020-21 made on the basis of 'business as usual.' - J.N. Maggo - R&D Needs for Energy Sector -September, 1996.

** Electricity generation capacity of 1745 TWH will be required to take care of T&D losses and Plant Load Factor. Corresponding installed generation capacity will be 378,000 MW at present day PLF and T&D loss level.

@ R.K. Sachdev - International Conference on Coal - June 1996.

Efficiency improvement is likely to make a substantial difference. Efficiency in generation and improvement in PLF will mean higher energy supply for the same primary input; and efficiency in use will reduce the demand of final energy. The combined effect in these—supply side and demand side—through proper management can be close to 1.5 to 2 per cent per year or about 25 to 30 per cent by the year 2020. The ultimate demand for installed power generation capacity can come down to the range of 275,000 MW to 300,000 MW; requiring additional capacities at an average rate of 10,000 MW per year upto year 2020.

Projections upto year 2020 have been made as the present trend of primary energy dominance will be replaced by a shift towards 'Bio-Nuc' era. Nuclear energy is expected to have installed capacity upto 20,000 MW, and biomass consumption in terms of fuelwood will be about 200 mt per year. However, other modern conversion technologies will be suitably adopted for biomass. Power generation from:(i) fuelwood incineration, (ii) agro-residue briquette gasification, and (iii) cogeneration in sugar mills; will have proven technologies. Most conservative estimate indicates that power generation from bio fuels will have an established 20,000 MW capacity, that too in dispersed fashion. "Bio-Nuc" era will have thrust on nuclear for base loads and bio fuels for dispersed loads. From 2020 onwards, additional capacities in power generation may be more from 'Bio-Nuc' combination than coal based power plants. However, in terms of percentage, Bio-Nuc may contribute to 12-14 per cent of power generation.

The emphasis on 'Bio-Nuc' era appears unavoidable as indigenous reserves of primary commercial sources like coal and petroleum would have dwindled to a very low level. On the other hand, thorium reserves can see India well into the year 2200; and biomass resources are renewable and would be available for a long time. The 25 years time span from now to 2020 provides sufficient breathing space to develop commercially viable technologies for bio fuels and safer options for nuclear waste disposal.

The nuclear power scientists and engineers have done well to utilize their lean period for 'rest, relief and training', a terminology used by the army for peace time operations. Their capacity to negotiate a cost effective deal with the Soviet Union for supply of two power plants of 1000 MW each; Nuclear Power Corporation promoting joint venture operations with power producing companies; and replacement of components of the Rawatbhata Atomic Power Plant are some of the examples as to how the Indian experts have learnt from their experience. The process of nuclear plant safety, waste disposal and nuclear radiation will need solutions, but these will evolve over time.

SECTORWISE PERFORMANCE DURING EIGHTH FIVE YEAR PLAN

Sectorwise performance is based on the review presented in the Economic Survey 1996-97—Government of India. This information has been added to give an idea about the current status of different energy sectors and some of the identified reasons for slippages from planned growth.

Power Sector

The power generation during 1995-96 at 380.1 billion kwh recorded a growth of 8.3 per cent over 1994-95. Thermal plants account for 74 per cent of total power generation and hydro electricity plants, 24 per cent, with the balance of 2 per cent is generated by nuclear plants. The installed capacity of power generation as on March 31, 1996 was 83,288 MW. During the first four years, of the Eighth Plan (1992-97), capacity addition has been 14,799 MW as against the Eighth

Plan target of 30,858 MW. About 12 per cent of the total capacity addition has been accounted for by hydro and the remaining by thermal units. Nearly half of the capacity addition occurred in the central sector. 42.5 per cent was accounted for by the State sector and the rest in the private sector (about 10,000 MW).

Several reasons have been identified for slippages in capacity creation in the central and State sector. These include paucity of funds, delay in placement of order for main plant and equipment, delay in supply of equipment by suppliers, procedural delay in land acquisition, non-resolution of inter-state disputes, problems due to disturbed site conditions, unresolved issues in fuel linkages, suspension of work due to contract failures and resettlement and rehabilitation problems.

The plant load factor has shown an upward improvement. Low capacity utilization of thermal power plants of the State Electricity Boards is largely due to deficiencies in management and operation, lack of proper maintenance and non-availability of coal of appropriate quality. There are wide inter-state variations; and the power load factors in the case of eastern and north-eastern region plants remain low.

An Action Plan drawn up by the Central Electricity Authority to improve the performance of the power sector, in a phased manner, has evolved short term, medium term, and long term plans. The Action Plan (given as Appendix-1) suggests that short term measures include: good housekeeping, especially maintenance of boilers, and optimal operation of regional grids. Significant improvement in PLF of thermal power stations can be effected through medium term measures like proper maintenance planning. In the long term the availability factor of the older thermal power plants can be improved by appropriate renovation and modernisation programmes.

The transmission and distribution losses which stood at 21.8 per cent in 1992-93 have come down marginally to 20.9 per cent in 1994-95. These figures still appear to be on the higher side when compared with the international average of less than 10 per cent of the advanced countries of the world. The general expert opinion is that 5-7 per cent of the present day T&D losses are basically due to power theft. If that is taken into consideration the figure for T&D losses will average around 15 per cent. That is no consolation, and consistent efforts are required to reduce the same; especially when one per cent reduction in T&D losses would result in saving of capacity of about 800 MW. Inherent losses in conductors and equipment can be brought down simply through system improvement schemes.

The financial health of State Electricity Boards and improvement in the operational performance continue to remain the more crucial issues in the power sector. The Electricity (Supply) Act, 1948 under Section 59 desires the State Electricity Boards to earn a minimum rate of return, R.O.R., of 3 per cent on their net fixed assets, after providing depreciation and interest charges. This provision was to become operative from the accounting year of 1985. But, the State Electricity Boards are yet to comply with this statutory stipulation. Whereas in 1992, the number of SEBs with a positive ROR were 11, there were only 6 such SEBs in 1994. Similarly as against 5 SEBs with a negative ROR in 1992-93, the number of such SEBs have increased to 9 in 1994-95. There were only 3 SEBs with a ROR of more than 3 per cent in 1994-95. The managerial and financial inefficiencies in the state sector utilities have, in turn, adversely affected future capacity addition and system improvement programme. On the one hand, the State Electricity Boards do not have enough resources to finance future programmes and, on the other hand, their ability to raise investible funds from alternate sources is limited due to their poor financial and commercial performance.

Some of the State Electricity Boards have introduced reforms and restructuring with a view to improving its physical and financial performance. The State of Orissa have provided the framework for reform envisaged through the enactment of a reform legislation, namely "The Orissa Reforms Act, 1995" made effective from April 1,1995. The Government of Orissa have initiated reforms of the state power sector, with substantial restructuring of the State Electricity Board to make the operation of the sector more efficient and financially viable. Orissa is the first State in the country to take up such an exercise. Under the Reform Programme, which is also being assisted by the World Bank through a US $ 350 million loan. The erstwhile Orissa State Electricity Board operating as a State monopoly combining functions, relating to generation, transmission and distribution of electricity has been replaced by separate corporate entities viz. Grid Corporation of Orissa, Orissa Hydro Power Corporation; and Orissa Power Generation Corporation. This restructuring has been effected through legislative changes made through the Orissa Electricity Reforms Act, 1995. A State Power Regulatory Commission with jurisdiction over issues relating to tariffs as well as licensing has also been set up under the Act. The reform effort envisages increasing private participation in the state power sector along with phased privatisation of the existing publicly owned entity. Emphasis has been placed on the reform of electricity distribution, with management of one zone (out of a total of four in the State), being privatised as an initial measure. Eventually all electricity distribution zones are proposed to be privatised on the basis of competitive bidding.

Oil and Natural Gas

The crude oil and natural gas liquids production during 1995-96 at 35.19 million tonnes showed a growth of 9 per cent over 1994-95. However a decline of 10.2 per cent in the first eight months of the financial year 1996-97 is a cause of concern. Some identified reasons for the decline include frequent power shut downs by Andhra Pradesh State Electricity Board, unexpected adverse behaviour of some of the reservoirs, environmental constraints and less base potential because of less number of wells available in Bombay High. This has resulted in higher import of crude oil and increased import bill.

The total refinery output during 1995-96 at 58.58 million tonnes showed an increase of 3.6 per cent over 1994-95.

Natural gas output at 22.31 billion cubic metre was 15.1 per cent higher than the production of 1994-95.Flaring of natural gas has now come down to 5 per cent, of which 2.5 per cent is the technical flaring. Systematic reduction efforts have considerably decreased the flaring and is likely to bring it down further.

The consumption of petroleum products at 74.72 million tonnes (including 2.15 million tonnes by the private sector imports) has shown an increase of 10.8 per cent over the previous year.

The gross import of crude oil and petroleum products during 1995-96, at 47.68 million tonnes, was 15.4 per cent higher than imports at 41.30 million tonnes, in 1994-95. The value of POL imports during the year was Rs. 24,095 crores as against Rs. 17,838 crores during the previous year. The gross import of POL during April—September 1996 was valued at Rs. 15,547 crores as against import valued at Rs. 9,876 crores during same period in 1996. This has been a major disturbing factor. Identified reasons for the same include increasing quantum of POL imports, depreciating rupee value and soaring prices in international markets. A major disturbing feature was the drop in indigenous production of crude oil.

The existing system of pricing of petroleum products is based on the report of the Oil Prices Committee, 1976, as amended by the Oil Costs Review Committee, 1984. The Government had set up a Reforms Committee (R-Group) with a view to develop a comprehensive long term perspective for the hydro carbon sector, and to meet the challenges that may emerge by the year 2010 and to study and suggest reforms in the petroleum sector. The report was submitted during the year and is under Government's active consideration.

Coal

Coal is one of the primary sources of energy accounting for about 67 per cent of total primary energy consumption in the country. Its production at 270.13 million tonnes in 1995-96 represented a growth of 6.5 per cent over the previous year. Given the present trends in production of coking coal and its high ash content, it is unlikely that imports of coking coal will go down in the immediate future. Imports of coal have been consistently increasing and reached a level of 13.22 million tonnes in 1995-96.

One of the major constraints on the profitability of coal sector is low productivity in underground mines. This has stagnated at an output per manshift of about 0.55 tonnes for the last two decades, despite massive investments made in mechanisation of underground mines. The underground mines employ 80 per cent of the manpower and contribute only 30 per cent of total output.

Two non-coking washeries with a capacity of 11 million tonnes per annum are likely to be commissioned in 1997. Four more washeries having a capacity of about 21 million tonnes per annum are to be built by Coal India Limited during the Ninth Five Year Plan period (1997-2002). That will reduce burden on coal transport sector.

Renewable Energy Sources

Contribution of renewable sources of energy is continuously increasing in the national sector. With technology development for conversion into improved final energy forms , the future applications may be on a large scale. A total capacity of 732 MW of wind generators has been established by the end of 1995-96, of which 684 MW has been installed in the private sector. Also a capacity of 121 MW has been created under the small hydro power projects, and 234 MW is under installation. The range under mini hydro power has been widened to include systems from 10 KW to 15 MW. Large joint sector projects are being launched for power generation through solar thermal route. Photovoltaic applications have established their need for small scale electricity needs and the remote areas.

A National Programme has also been launched since June 1995 for proper treatment of urban and industrial wastes and resultant recovery of energy from these sources. With the launching of this programme an era of modern technologies has been initiated . Other applications of bio energy include promotion of fast growing species, gasifier systems for power generation, briquetting of agro residues, biogas digesters and improved chulhas.

Exciting innovations have been introduced in the solar thermal sector. Earlier efforts at technology development has resulted in commercial development of market for solar water heaters, solar cookers etc. Commercial banks are also launching schemes for funding such devices.

ENERGY SECTOR INVESTMENT

Plan Expenditure

The development of commercial forms of energy is highly capital intensive and large investments are required for meeting the rapidly growing demand for these forms of energy. The relative share of expenditure on energy sector in the total Plan expenditure during successive Five Year Plans is shown in Table 1.13.

TABLE 1.13: Plan Expenditure and the Share of Energy Sector

Plan	Period	Plan Expend. (Rs.Cr.)	Share of Energy Sector (per cent)				
			Power	Oil/Gas	Coal	NRSE	Total
First	1951-56	1,960	-	-	-	-	19.7
Second	1956-61	4,670	9.7	0.8	1.9	-	12.4
Third	1961-66	8.580	14.6	2.6	1.3	-	18.5
Fourth	1969-74	15,780	18.6	1.9	0.7	-	21.2
Fifth	1974-79	39,430	18.7	3.6	2.9	-	25.2
Sixth	1980-85	109,290	16.7	7.8	3.5	0.1	28.1
Seventh	1985-90	220,220	17.4	7.3	3.2	0.3	28.2
Eighth	1992-97	434,100	18.4	5.5	2.4	0.2	26.5

Ref. Economic Survey, Government of India

In the Eighth Five Year Plan, an outlay of Rs 434,100 crores was provided for different sectors of the economy. The share of energy sector was kept at 26.5 per cent of the total public sector outlay. The power sector was allocated a share of 18.4 per cent.

Private Sector Investment

The Government has laid down policy for private participation in the energy sector. Procedures are now evolving as the policy is being implemented. Details of various steps taken so far are given below:

Power

The Government has laid down norms and procedures for private sector participation so as to meet the growing demand in the power sector. The policy with the object of mobilising additional resources for capacity addition in power generation and distribution had been formulated in 1991, and is currently under implementation. The response to the policy has been encouraging, though the private investment in real terms and its impact in the power generation is to be seen on the ground. As a first step, the Government had decided to support eight fast track projects under Government of India's Counter-Guarantee Scheme. Three of these projects are at different stages of construction; one project is under renegotiation; for one more project power purchase agreement has been finalised and the other three are under different stages of PPF finalisation. During the first five years there has been no increase in the power generation through private sector. Status of fast track projects is given at Appendix-2.

Under the current structure, the Independent Power Producers have to sell power to State Electricity Boards who will undertake its distribution. Since most SEBs are financially distressed, the security package for an IPP is considered crucial for the success of a power project. The Government's counter-guarantees to the State guarantees was considered necessary for the initial stage to attract private capital into the power sector.

To facilitate setting up of large size thermal/hydro electric plants in the country, and to derive economies of scale, projects having a capacity of 1000 MW or above and supplying power to more than one State are being considered as mega projects and receive separate treatment. A notification issued in October 1995 enhanced the limit of capital expenditure on schemes requiring concurrence of Central Electricity Authority from Rs 100 crores to Rs 400 crores in the case of generating companies. This limit has been further enhanced to Rs. 1000 crores in September 1996. The State governments have expressed opinion to raise the limit to 1000 MW installed capacity.

Oil and Natural Gas

The Government has cautiously opened the oil and natural gas sector for private investment. Six rounds of bids for exploration in order to increase reserve accretion have been carried out. Government has awarded contracts for 35 exploration blocks. Attractive fiscal incentive package has been proposed for the exploration in deep water and frontier areas, which have high risk but with a potential of high hydrocarbon reserve.

The Government has also offered discovered medium sized oil and gas fields for development under joint venture arrangement with 40 per cent participating interest for national oil companies, including small sized fields which have been awarded to private companies. Government has so far awarded 30 fields for development.

In the refining sector Government has issued letter of intent to 12 private sector parties for a total refining capacity of 70.3 million metric tonnes per annum of petroleum products by 2001-02. Seven of these units are export oriented. Transportation of petroleum products and natural gas is preferred through the pipelines. As the investments required are heavy the government has approved setting up a holding company and subsidiary joint sector companies for expeditious implementation of new pipeline projects.

Coal

In order to encourage private investment in the coal sector, Coal Mines Nationalisation Act, 1973, was amended with effect from June 1993 for operation of captive coal mines by companies engaged in the production of iron and steel, power generation and for washing of coal in the private sector. Production of cement has also been notified as an end use with effect from March 15, 1996 for the purpose of captive coal mining. Imports of coal under Open General Licence is allowed. Captive coal mining blocks have been identified for 14 power generating companies/electricity boards and five companies in the iron and steel sector.

APPENDIX-1

Common Minimum National Action Plan for Power, 1996

* The Government would soon finalise a National Energy Policy. Each State/Union Territory shall set up an independent State Electricity Regulatory Commission (SERC). To start with such SERCs will undertake only tariff fixation.
* The Union Government will set up a Central Electricity Regulatory Commission (CERC) which will set the bulk tariffs for all Central generating and transmission utilities.
* Determination of retail tariffs, including wheeling charges etc., will be decided by SERCs which will ensure a minimum overall 3 per cent rate of return to each utility.
* Cross-subsidisation between categories of consumers may be allowed by SERCs. No sector shall, however, pay less than 50 per cent of the average cost of supply (cost of generation plus transmission and distribution). Tariffs for agricultural sector will not be less than fifty paise per kwh to be brought to 50 per cent of the average cost in not more than three years.
* The Central Government would make a comprehensive review of the role of the CEA. Techno-economic approval of competitively bid power projects will be simplified and CEA shall not be concerned with capital cost, tariff and other commercial aspects of the project. However, CEA's appraisal will continue in respect of planning and other related matters.
* States will allow maximum possible autonomy to the SEBs. SEBs would professionalise their technical inventory manpower and project management practices.
* State Governments agree to a gradual programme of private sector participation in distribution of electricity.
* State Governments will encourage co-generation/captive power plants. To facilitate evacuation of power from these plants to the grids, States shall formulate clear and transparent policies for purchase of power and wheeling charges which provide fair returns to the Cogeneration/Captive power plant owners.
* The role of FIPB will be minimised by putting as many projects on the automatic clearance route as feasible.
* Government of India will carry out necessary amendments in the relevant Acts/Rules to allow private participation in transmission.
* A national policy on hydro power development will be evolved by the Central Government.

APPENDIX-2

Status of Eight Fast Track IPPs approved for GOI Counter Guarantees

S. No.	Name of Project/ Promoters	Capacity (MW)	Provisional Cost (Rs.Crore)	Latest Position (as on January 30, 1997)
1.	Dabhol CCGT, Phase i/ Maharashtra (M/s Dabhol Power Company-promoted by Enron, Bechtel and General Electric)	740	2912	Project under construction
2.	Jegurupadu CCGT/Andhra Pradesh (M/s GVK Reddy)	216	816	Project under construction. First three units of the project commissioned.
3.	Godavari CCGT/Andhra Pradesh (M/s Spectrum Power Generation)	208	748	Project under construction. One unit commissioned.
4.	Ib Valley TPS, Units 3 & 4/Orissa Counter Guarantee issued. (AES Transpower, USA)	420	1994	Project being renegotiated by Government of Orissa/ OSEB.
5.	Neyveli (Zero Unit)/ Tamil Nadu (ST-CMS Electric Co.)	250	1200	Power Purchase Agreement (PPA) finalised.
6.	Managalore TPS/Karnataka, Mangalore Power Co. (a subsidiary of Cogentrix Inc.USA)	1000	3948	PPA being finalised.
7.	Visakhapatnam TPS/Andhra Pradesh (Ashok Leyland and M/s National Power plc., UK)	1040	4797	PPA being finalised.
8.	Bhadravati TPS/Maharashtra (M/s Nippon Denro Ispat Ltd.)	1072	4362	PPA being finalised.

2

STRATEGY FOR RURAL ENERGY SUPPLY IN INDIA

INTRODUCTION

The economic development in India under recent thrust on market oriented approach has opened avenues for larger involvement of private sector in creating necessary infrastructure. The economic reforms of the past five years have got around a recovery in the growth of production sector and employment and has resulted in far-reaching changes in agriculture, industry, finance, capital markets and the tax structure. The Economic Survey 1995-96 published by the Government of India in July 1996 has focused on the major challenges and policy priorities confronting the Indian economy. As the government's approach is to reduce subsidies there is an urgent need to create more jobs in the rural areas and improve rural income generation by creating necessary infrastructure.

Two-thirds of India's labour is employed in agriculture, and it is crucially important to broad-base agricultural growth for raising rural living standards, ensuring basic food security for the nation and strengthening the basic markets for industrial and service sectors.

Poverty alleviation programmes have received an overall support at various national and international conferences that took place during 1996. These include the International Conference on Habitat, the World Solar Summit and the Preparatory Meetings of the forthcoming World Food Summit. There was a general agreement on the need for an overall improvement of the rural sector.

The strategy for energy development forms an integral part of

Paper presented to the Meeting of WEC Study Committee on Energy Supply in Developing Countries, organised by the World Energy Council on October 7, 1996 at Kusadasi (Turkey)

the overall economic development strategy in India. Efficient use of resources and long term sustainability are two important objectives of the economic planning. The concept of sustainability takes account not only of natural resources and ecological balance, but also economic equity and self-reliance. Any strategy for energy planning, therefore, is dovetailed with these broad objectives.

Indian economy continues to be primarily based on agriculture, even though its share in the GDP at constant prices has been declining over time. The share of agriculture sector in the GDP has declined from 39.7 per cent in 1970 to 29.5 per cent in 1993; whereas that of the industries sector has gone up from 17.4 per cent to 22 per cent during the same period. The average rate of growth in GDP, which was about 3.5 per cent up to the 70s, has averaged more than 5.60 per cent during the 90s. This trend in GDP growth is to improve in the years to come. Overall economic growth of GDP at factor cost, accelerated to 7 per cent in 1995-96. The growth rate was not due to exceptional agricultural growth, the high growth rate was mainly due to the 11.7 per cent growth of industry and construction sector and the 7 per cent growth of the service sector. The overall economic growth during the first four years of the Eighth Five Year Plan (1992-97) averaged 5.9 per cent per annum, a rate significantly higher than the Plan target of 5.6 per cent.

The situation is building up to give prominence to rural sector in India. Consequently, rural energy is given due importance in India's economy.

RURAL ENERGY PLANNING

Rural Energy is basically reflected for the minimum essential needs of cooking energy. At least that is how it was started in India. Firewood, agro-residue and cow dung are considered the basic fuel sources. Only firewood could be planned and quantified to a certain extent. Primary data are available from field surveys, conducted by the National Sample Survey Organisation (NSSO), National Council of Applied Economic Research (NCAER)and other Research Institutions and individuals.

J.N.Maggo, in his presentation to the Fuelwood Planning Course of RWEDP in 1995 summed up India's efforts at rural energy planning as follows:

> "The methodology of estimation of household energy demand was first given by the Energy Survey of India Committee (ESIC)

in 1965. Based on the results of the sample survey carried out by certain agencies, the ESIC adopted a norm of 0.39 tonnes of coal replacement (tcr) of energy consumption requirement per person per year in the urban households and 0.38 tonnes of coal replacement of per capita energy consumption per annum in the rural areas. After so determining the total energy requirement in the household sector, the projected contribution of commercial sources of energy was subtracted from it to arrive at the share of non-commercial fuels. The same was then apportioned among fuelwood, agricultural waste and cow dung in the proportion of 65 : 20 : 15. This is only to indicate that the available estimates of non-commercial energy consumption in the household sector yield only rough estimates rather than precise estimates. This methodology continued to be adopted till the mid-80s. The Advisory Board on Energy adopted the concept of useful energy requirement per person in the household sector. In the report on 'A Perspective on Energy Demand and Supply in India in 2004-05' published in 1985, the ABE adopted a norm of 650 kilocalories of useful energy requirement per person per day for cooking and space heating in the year 2004-05. The fuel mix for the year under consideration was then worked out on the basis of population, the rural-urban distribution of population etc. The fuel mix adopted for the exercise was as per the Domestic Fuel Survey of the NCAER for the year 1978-79. Later in 1986 the Energy Demand Screening Group (EDSG), set up by the Planning Commission, modified the consumption requirement of useful energy of 520 kilocalories per person per day in the rural areas. Other studies carried out later, like the Sectoral Energy Demand in India, also estimated the useful energy requirement in 2009-10 as 641 kilocalories per person per day."

The Sixth Five Year Plan (1980-85) recognised fuelwood as the most important non-commercial fuel which was becoming increasingly scarce and also getting commercialised. Energy forestry was also recognised as an important component of the energy strategy as it was felt that the rural communities will continue to depend on firewood for several decades to come. A large programme of fuel and farm forestry was taken up in the plan. Also, an Integrated Rural Energy Programme (IREP) was initiated in the Sixth Plan on a pilot scale with an objective of providing a mix of energy options for meeting the energy needs of the rural areas on an area-wise basis in the most cost-effective manner. To implement this programme, the Planning

Commission set up a separate Rural Energy Division. A Rural Energy Planning Exercise was commissioned by the Planning Commission to develop the design and approach of planning and implementing area-based block-level integrated rural energy projects.

In a number of studies conducted by IAAS, it was observed that the rural domestic energy consumption in the final form varied from 420 kilocalories to 620 kilocalories per capita per day, depending on the status of the family and the use, whether only cooking or cooking and water heating.

The Advisory Board on Energy brought out a report 'Towards a Perspective on Energy Demand and Supply in India in 2004-05' in 1985. The main conclusions of the report pertaining to supply options, analysed Statewise, in the case of firewood were that at least 60 million hectares of land (comprising open/degraded forest land, barren and unculturable land, culturable wasteland, land other than current fallows and farm perimeter land) might be available for raising fuelwood/fodder plantations. Given the acute pressure on land and the other potential uses it could be put to, not more than half of the land area thus identified could actually be used for plantation purposes. The study brought out that the productivity of the Indian forest land is quite low and need to be increased.

The Government of India also set up a Wastelands Development Board in 1981 to explore the possibility of greening wastelands for energy plantation. Though 92 mha are reported available, only about 20 mha can be economically exploited. Expert Consultations, including those under FAO and IAAS, have indicated that energy plantations with long term yields of 4 to 5 tonnes per ha per year can be certainly grown on various wastelands and village lands.

Yields of upto 40 tonnes per ha have been recorded in test fields. Subsequent policy decisions to separate community based/farm forestry energy plantations from large forest lands has been found a good option. Various studies in places like Haryana, Kerala and Western Uttar Pradesh have indicated that the local availability of fuelwood from village woodlots has resulted in stabilising the local market price of fuelwood.

PRESENT RURAL ENERGY PROGRAMMES

Rural Electrification Programme

The Five Year Plans (upto 1970) considered agricultural pump energisation as an important thrust area. It was initially envisaged to provide electricity as a social amenity to rural areas and was confined to a few States. Subsequently, it was extended to cover all the States. The importance of this programme was especially recognised during the drought in mid-60s when lift irrigation had to be resorted on a large scale to several subsistence crops. This programme was subsequently integrated with Minimum Needs Programme for the rural poor.

The Programme was further strengthened by the formation of Rural Electrification Corporation in 1969, which now provides over 90 per cent of the funds to the State Electricity Boards for rural electrification, as concessional loan. In addition to energisation of pump sets, emphasis was also laid on promotion of industrial load in villages by extending electricity connections for setting up low tension cottage/home based industries.

The present thrust under the Programme is as follows:

i Improvement in the system of power distribution in the electrified villages to ensure quality and reliability of supply.

ii. Load development activities in electrified villages to promote the use of electricity for productive purposes other than irrigation, so as to provide wider opportunities for the rural communities for gainful employment and to improve their incomes. In this context rural and agro-based industries will be specifically promoted in coordination with the rural electrification programme. Present growth of rural industries average around one small industry per village in backward areas and about two per village in developed areas with power capacity ranging from 5 H.P. to 25 H.P. Each industry employs around 1.45 persons on an average.

iii. Development of co-operatives, preferably operated by the panchayats and other local bodies for distribution of electricity in the rural areas so that there is an improvement in collection of dues, reduction in thefts of electricity and T&D losses.

Integrated Rural Energy Programme

Besides the rural electrification programme, a new approach of integrated rural energy systems was also promoted. The Planning Commission initially started an Integrated Rural Energy Planning Programme (IREPP) which was taken up as a plan scheme in the Seventh Five Year Plan.

A need was felt to understand patterns of energy consumption at the micro-level, through decentralised energy planning exercise, in order to provide sustainable and affordable supply of energy sources for meeting the growing energy needs of the rural population. While locally available renewable energy sources based on non-conventional technologies have been projected to play an increasingly important role in the coming years in meeting rural energy needs, contribution of commercial energy for productive and income generating activities was also realised.

The Government of India decided to meet all expenses for the institutional arrangements. The Central Plan component was utilised for setting up the institutional mechanisms in the State/UT through funds for professional and support staff in the IREP cells at the State level and in the selected districts/blocks, as well as for the training. The State component of IREP funds was utilised for project preparation and implementation and financial incentives for the promotion of rural energy technologies, as part of the block level IREP projects.

The training component involved national, regional, and state level academic professional institutions, besides technical colleges/ universities which were provided technical and financial support for conducting training courses on various aspects of the IREP programme. The setting up of one national and four regional training-cum-R&D centres was also taken up under the programme in the Seventh Plan. The first national level training-cum-R&D centre (Centre for Integrated Rural Energy Planning, CIREP) was set up at Bakoli village in Delhi in cooperation with Delhi Administration and with the technical and financial support of the Planning Commission under the centrally sponsored scheme for IREP. The other four regional centres are located in Lucknow (UP), Bangalore (Karnataka), Kheda (Gujarat), and Shillong. Technical back-up units were set up in selected state level institutions. Also, district level IREP technical back-up units were set up through ITI/Polytechnics in selected IREP blocks to provide technical assistance especially in the selection, operation and maintenance and follow up of applied R&D in rural energy technologies in the IREP projects.

Environmental Consideration

Another aspect that needs to be considered is that the inefficient burning of biomass emits smoke causing health hazard to beneficiaries who are largely in the lower income groups of the rural and urban areas. In the normal course, the major objective of energy planning should be either to replace these traditional fuels completely or introduce cleaner and more efficient technologies for the conversion of biomass into useful energy for cooking. The latter option will call for major investments as well as technology upgradation. In the case of dung, biogas is a well established technology. If the traditional fuels are to be replaced completely, it will require either kerosene or coal-based fuels. Taking into account the relative calorific values as well as the end-use efficiencies, the requirement of equivalent quantitities of coal and kerosene for replacing these fuels have been estimated. In other words, to protect the rural and the urban poor from explore to excessive emission of smoke, it is necessary to step up fossil fuel supplies, either petroleum products or coal or both.

Consumption of Cooking Energy and Expenditure

Indian Association for the Advancement of Science, IAAS, conducted a survey of dietary energy and cooking energy in different parts of the country during 1992. Details of the same are given below:

A comparison on the basis of dietary energy intake alone may disorient the projection. Comparison was made on the basis of total expenditure on the cooked foodstuff, i.e., on the dietary energy intake and on the cooking energy source. It was observed that in the tribal, desert and the hilly region, firewood and dung cakes were the cooking energy source and in case of coastal area and plains it was dung cakes and LPG. Reasons are simple that in cases where the people are poor or remote from the urban centre LPG supply was not feasible. On the other hand in case of Singhu which was a part of a large metropolitan city of Delhi, LPG was abundantly available and the income levels are also such that LPG can be afforded. Another observation was that the firewood in the metropolitan city cost Rs. 1.50 per kg and therefore use of firewood was non-existent and only dung cakes were being used.

At the time of survey, there was no commercial value attached to firewood taken out of forests by unauthorised tree felling; and also no commercial value was assigned to dung cakes as the cow dung was freely available and the labour for making cakes out of it was from the family itself. Therefore, there was no effect of cooking energy cost in the tribal, desert and hilly villages but there was a small component of 14 paise per capita per day alone in case of Singhu village. If normative price of the cooking energy source was taken then it was observed that in the tribal village of Peepalwas cooking energy costs Rs.1.22 per capita per day; in desert village of Madni the cooking energy costs Re. 0.79 per capita per day; in case of hilly village of

TABLE 2.1: Consumption of Cooking Energy and Expenditure

	Site	Cooking		Lighting		Total Expenditure (Rs)
		Energy (Kcal)	Expenditure (Rs)	Energy (Kcal)	Expenditure (Rs)	
1.	Singhu	598	1.24 (0.14)	391	0.45	1.69 (0.59)
2.	Niglat Nainital	523	1.10 (-)	189	0.13	1.23 (0.13)
3.	Madni Sikar	373	0.79 (-)	107	0.09	0.88 (0.79)
4.	Peepalwas Udaipur	581	1.22 (-)	266	0.09	1.31 (0.09)
5.	Rajpuriya Dhar	418	0.58 (-)	292	0.29	0.87 (0.29)
6.	Pallip- puram Ernakulam	470	0.94 (0.19)	293	0.33	1.27 (0.52)

Niglat the cooking energy costs Rs.1.10; and in the case of plains village of Singhu the cooking energy cost is Rs.1.24 per capita per day. Refer Table 2.1.

Cooking Energy sources used are firewood and dung cake only in case of 2, 3 and 4; and dung cake and LPG is used in case of 1. Users in 1 pay only for LPG. Figures in brackets indicate actual paid and others represent normative market price of source use.

THE FUTURE COURSE

The future course of rural energy consumption will depend on government's policy. Under present conditions, the rural India will witness emergence of rural based industrial units and their increasing contribution to nation's economy.

Rural Industry

Benefit of industrialisation can be distributed to the rural areas by setting up or promoting small scale but relevant industries. Apart from basic infrastructure elements such as energy and transport, the necessary technological infrastructure must be created in terms of entrepreneurial capability, growth of skills and the like. Despite its complexity greater industrial dispersal may have to be undertaken leading to greater income and employment in the rural communities.

Nature and pattern of rural industrial activity could cover a wide range, including the following:

i. Repairs and servicing of transport and agricultural equipment;
ii. Establishment of low capital and labour intensive ancillary industries;
iii. Setting up processing units for both perishable and non-perishable local produce;
iv. Production of simple agricultural equipment, such as farm implements and pumps;
v. Manufacture of products to meet local needs, such as building materials, cloth and shoes.

Rural Energy Supply Options

Realising that the conventional energy supply systems that have evolved in conjunction with large scale, highly concentrated production and distribution units have often proved to be inadequate for the needs of the rural population in India. They have often failed to yield adequate electrical power in relation to the micro-economic needs; and have usually not met the total specialised demands of agriculture or of rural and small scale industries.

When properly used, small and perhaps intermittent amounts of energy can be of critical value to the rural economy. The energy planning for rural areas involves development of self sufficient local systems. In some cases, such a programme might even involve suitable phasing and postponement of centralised rural electrification schemes in order to release more resources to implement more quickly maturing decentralised schemes.

Objective

IAAS and other Expert Groups have realised that the objective of Rural Energy Planning in India, under present trend of economic growth, should be:

i. To increase the energy availability to the rural areas so as to reduce human drudgery, especially of women, such as the long hours and physical effort required for gathering fuel, fetching water and carrying heavy loads;
ii. To give spur to the economically productive activities necessary for rural development; and

iii. To create an organisational framework to initiate, introduce, implement and maintain conventional and new energy technologies.

Strategy

The strategy to meet rural energy needs for the rural and remote area communities should have the following components:

i. Harmony between planning for rural energy needs and ongoing activities of rural development agencies.
ii. Planning for energy needs in an integrated fashion so that the community needs and the income generation needs are met in a balanced fashion; and implementing programme and projects accordingly.
iii. Planning to meet energy needs of a cluster of villages, to achieve cost effectiveness; and implementing them through community based infrastructure.

Benefits

Benefits from the outcome of such a strategy are expected to be as follows:

i. Solution to employment generation, as more self-employment programmes will be supported in rural areas.
ii. Rural-urban migration to be restricted as employment opportunities in the rural areas will be available for larger numbers.
iii. Sustainable development of the area possible as most of the locally available resources will be utilised with total participation of local communities.
iv. Higher incomes due to value addition at source and thereby better prosperity for the people of the area.

CONCLUSION

A decision has been taken to allow smooth transition to urban-rural continuum, as projected at Habitat-II. Whereas rural urban migration will not stop, wealth from urban areas will be transferred to the rural areas through government and private agencies and individuals. But their sustenance will depend on affordable core infrastructure, including energy. Energy supply in future in the rural areas in India will be dictated by market forces, that are evolving now. The booming small scale industries sector and the service sector will dictate rural energy mixes and rural energy supply options. Government will not be the only sector to fulfil these needs. The economic growth will itself set a pattern for fulfilling rural domestic energy needs. It would not be a surprise if the domestic energy needs in the rural areas match or surpass the urban area needs. That may take time, but the process has already set in motion.

A shift towards liquid fuels and electricity is unavoidable. That does not necessarily mean supply though commercial, centralised sources of energy. That will, in all possibility, come through environment friendly small scale bioenergy based conversion devices, operated on a large scale by young entrepreneurs in rural areas.

The future in India will see a paradigm shift in energy consumption patterns. Commercial sources will have a larger role to play in rural areas because of transfer of wealth and prosperity and the urban areas will see higher consumption of fuelwood based conversion devices, obviously more efficient than now.

3

ENERGY CONSERVATION PERSPECTIVES FOR INDIA

INTRODUCTION

India's population was 844 million as on March 1,1991. About 26 per cent of the population lives in urban areas. The population of the four metropolitan cities of Delhi, Bombay, Calcutta and Madras account for nearly 4.6 per cent of the total population of the country. The population is estimated to cross the 930 million mark by the end of the current year (1995).

Indian economy continues to be primarily based on agriculture even though its share in the GDP at constant prices has been declining over time. The share of agriculture sector in GDP has declined from 39.7 per cent in 1970 to 29.5 per cent in 1993 whereas that of the industry sector has gone up from 17.4 per cent to 22 per cent during the same period. The average rate of growth in GDP, which was around 3.5 in the 70s, has averaged more than 5.6 per cent during the 80s. This trend in GDP growth is likely to continue in future.

The coefficient of elasticity of final commercial energy consumption with respect to GDP at constant prices has been 1.14 during the period 1970-1993. The elasticity coefficients for coal, electricity and refined petroleum products for the corresponding period are 0.54, 1.82 and 1.43 respectively.

As far as energy intensity of the economy is concerned, the primary commercial energy intensity has increased from 1841 KJ/ Rupee (at constant prices) in 1970 to 2445 KJ/Rupee in 1993 whereas the final commercial energy intensity has increased from 1411 KJ/

Keynote Address at National Workshop on Energy Conservation, organised by the Rajiv Gandhi Foundation and The Institution of Engineers on September 8-9,1995 at New Delhi.

Rupee to 1605 KJ/Rupee during the same priod.

ENERGY CONSERVATION EFFORTS IN INDIA

The Government of India took a few major energy conservation initiatives as listed in Annexure I. The fiscal measures adopted are listed in Annexure II.

The Advisory Board on Energy, ABE, in February 1983, recommended setting up of a well structured and funded energy conservation organisation as early as possible to act as a nodal point. ABE, through its high powered structure had created a significant impact on the need for energy conservation. Also the Chairman of the ABE, being an eminent person, carried weight with the government and could deal directly at the highest possible level.

The Inter-Ministerial Working Group Report 1983 has made a series of recommendations on the organisational infrastructure required to support a comprehensive energy conservation programme.

The government founded the Energy Conservation Cell in the Department of Power, with the responsibility of carrying out energy conservation activities. The government also appointed an Adviser (Energy Conservation) in the Cabinet Secretariate, who became a nodal point for the energy conservation activity from June 1990.

The Department of Power in 1989 established the Energy Management Centre (EMC), an autonomous organisation, to assist in an energy conservation programme with the foreign assistance from the EEC, World Bank, UNDP and other agencies.

Following are some selected efforts initiated by various organisations:

NATIONAL PRODUCTIVITY COUNCIL

The National Productivity Council was the pioneer in energy auditing of the industrial sector. Two hundred industrial units covering 12 sectors indicated a potential for saving energy upto 25 per cent in the industrial sector. An estimate was also made that an investment of Rs. 51,400 million in energy conservation scheme would save a capital expenditure of Rs.79,800 million on energy production besides reducing energy costs of the consuming sectors by avoiding a recurring annual energy bill of Rs.37,190 million. A follow up study by NPC in 1985 of 178 units in 12 sectors covered in the earlier study revealed that 70 units (40%) had implemented energy conservation programme successfully, while 29 units (16%) were yet to start energy conservation programme. The pace of implementation has generally been slow and the energy savings achieved were Rs.102 million per annum against a theoretical potential of Rs. 830 million, indicating that 12.3 per cent of the saving potential has been achieved in these units.

As a component of the Indian Power Sector Assessment Study, conducted by the National Productivity Council in 1991, under the World Bank and bilateral donor agencies, it was estimated that 20 per cent saving in demand projections are technically cost effective. The estimate based on the potential and applicability of the 27 energy conservation opportunities, such as high efficiency motors, compact fluorescent lamps etc., can realise a saving of capital cost of Rs.43,000 to Rs.69,000 crores depending on the low cost and high cost scenario projected by the CEA in India Power Sector by 2004-05. This would mean a reduction in peak demand of about 38,000 MW and 61,000 MW respectively. Technologies for the energy efficiency equipment, products and processes have to be developed or technology transferred or imported to realise this potential.

FEDERATION OF INDIAN CHAMBERS OF COMMERCE AND INDUSTRIES

Another survey by the Federation of Indian Chambers of Commerce and Industries (FICCI) in 1988, carried out in 341 units, revealed that several industries were able to realise savings as shown in Table 3.1.

TABLE 3.1: Energy Saving Achieved By Respondents

	No.of Units	% of total Respondents
Upto 5%	128	37.54
5 - 10%	72	21.11
10 - 15%	34	9.97
15 and above	43	12.81

Within the country itself one perceives wide variations in energy consumption within a sub-sector, for example in the steel sector, the consumption varies from 8.7—15G Cal/tonne of crude steel. The figure for developed countries is 4–6 G Cal/tonne of crude steel. The energy consumption trends in SAIL plants indicate a lowering in consumption with advancing years (See Table 3.2).

TABLE 3.2: Energy Consumption in SAIL Plants

Year	Energy Consumption G Cal/tonne Crude Steel
1984	11.2
1987-88	10.7
1989-90	10.2

FICCI had subsequently carried out a number of other similar studies for other sub-sectors of industry.

ASSOCIATED CHAMBERS OF COMMERCE AND INDUSTRIES

Associated Chambers of Commerce and Industries (ASSOCHAM) in 1991 covered the sub sectors of chemicals, cement, engineering, automobiles, polymers and rubbers, metals, textiles, leather and food industries. The study revealed that, in general, large industries were found to be well organised and well informed in energy audit subsidy schemes; and equipment financing for energy conservation schemes. Such awareness was not generally evident with medium and small scale industries. Several of the industrial units had undertaken energy audit with professional assistance from outside consultants. The percentage share of energy cost in the total cost varied from a low of 0.36 per cent in Farm Equipment Limited to as high as 65 per cent in Chemplast.

Following variations in terms of cost for energy consumption were observed:

Engineering industry (including automobile)	0.36-2.1%
Cement	21-36%
Plastics	6.00-30.96%
Textiles	0.33-10.4%
Metals	3.34-7.3%

ENERGY CONSERVATION AND EFFICIENCY IMPROVEMENT—THE OBVIOUS ROUTE

Energy conservation is the only route that can get better mileage out of the available resources. The need is to consider the possibility of evolving an appropriate strategy for energy conservation measures in the country. It is, no dubt relevant to study the approaches followed in different countries, but it is equally important to understand the impact felt so far in economic terms because of the conservation measures adopted elsewhere. That will put our approach in the right perspective.

There are also general and obvious points that can get overlooked

(i) Energy intensity generally falls because productivity gains have raised the value added per unit of energy and because of structural shifts in the economy. These changes may have little or nothing directly to do with deliberate efforts to raise energy efficiency.

(ii) Similarly, technological innovations may well have as a by-product the reduction of energy usage, without this having been a deliberate objective. Hence the "natural" turnover of capital stock usually leads to a reduction of energy usage. In the future, this situation may change and accelerate due to shifts in energy policy and prices. In the past there has been little experience of the impact of measures to raise energy efficiency.

It is fortunate that innovations and, above all, turnover of capital stock across the broad economic spectrum result in declining energy intensity. An International Energy Agency Report—"Energy Efficiency and the Environment", 1991, (p.48) warned: "The progress of energy efficiency with the demand trends examined, needs to be evaluated with care because it is inter wined with so many other factors affecting demand. An energy efficiency improvement is defined as any action by a producer or consumer of energy that reduces the use of energy without affecting the level of service provided".

BARRIERS TO ENERGY CONSERVATION EFFORTS

It is relevant to consider initially the various barriers in terms of institutional, financial and technical arrangements as narrated below (categorisation as per WEC).

Institutional

(i) poor information about, and lack of experience, in the energy sector.

(ii) ignorance of what is available in terms of technology and know-how to promote efficiency and proper maintenance;

(iii) bias in favour of encouraging energy provision and usages without similar bias towards efficiency;

(iv) unwillingness to press for full-cost pricing, in developing countries usually for understandable social and political reasons;
(v) opposition from politicians and public revenue officials to the imposition of earmarked taxes to promote energy efficiency.
(vi) poor linkages between utilities and users, and between purchasers and users;
(vii) intellectual property rights, which are intended to reflect the efforts and costs put into developing new ideas, products and processes;
(viii) unwillingness to promote public information compaigns and expose inefficiencies;
(ix) ineffective regulation and labelling; and
(x) electoral concerns.

Financial and Economic

(i) perceptions that costs outweigh benefits (despite the fact that payback periods are often in months rather than years and that some efficiency improvements can be achieved at close to "zero cost" — even sometimes with net savings;
(ii) difficulty in identifying and calculating savings with precision;
(iii) value of, and returns on, existing assets;
(iv) financial accounting and budgeting methods;
(v) low energy prices, subsidies, and regimes which encourage marginal demand at minimal cost;
(vi) threshold level of energy and cost savings (below which investments do not seem worth making);
(vii) perceived high capital cost of energy efficient technologies as a barrier for poor, heavily indebted countries.

Technical

(i) non-availability of more advanced technology;
(ii) non-availability of new materials and other substitutes;
(iii) infrastructure limitations;
(iv) cultural norms;
(v) tendency toward over-sizing/low capacity utilization;
(vi) lack of research, development and demonstration;
(vii) lack of technical and managerial expertise;
(viii) poor maintenance and lack of spare parts.

IDENTIFIED POTENTIAL AREAS FOR ENERGY CONSERVATION

Following areas have been identified where specific energy conservation measures can have favourable results:

Industrial Sector

Following specific industrial sub-sectors have the largest scope for energy conservation and efficiency improvement without involving heavy modernisation costs:

- (i) Processing industries;
- (ii) Petroleum refining;
- (iii) Aluminium;
- (iv) Paper;
- (v) Cement;
- (vi) Steel;
- (vii) Sugar; and
- (viii) Copper.

Transport Sector

This will include automobile manufacturing industry as well as road vehicle and railway coach maintenance with emphasis on two wheelers in road vehicles.

Agriculture Sector

Emphasis is to be placed on diesel engines and pump sets and their accessories.

Domestic Sector

Emphasis is to be placed on good housekeeping.

Power Generation and distribution sector

Power generation and distribution sector has been identified as a major loss incurring sector. Different steps can be taken which will have a combination of managerial inputs and technological inputs to bring in higher economic returns.

SECTORWISE ISSUES SUPPORTING CONSERVATION

Industrial Sector

The energy consumption pattern observed in industry during last three years has shown a lowering in energy intensity per unit of industrial production. This is a very positive sign for India where investments in stocks are heavy and foreign investment and private investments are not coming forth. If a lead is to be taken from recent observations then one has to really appreciate the efforts initiated about ten years back. It is time to intensify efforts to penetrate medium scale and small scale sectors of the industry. It has to be noted that whereas large industries depend for modernisation mostly on the imported equipment, and thereby, they have better access to the latest energy efficient technologies from the international market; the medium and small scale industries largely depend for their equipment and plant on the local manufacturers. In such a case, unless the manufacturers of the equipment and machines are given certain incentives they may not be motivated to produce energy efficient equipment and machinery; and correspondingly the energy efficiency results may not be very encouraging from the medium and small scale sectors of the industry. It will be necessary, therefore, to think in terms of creating an awareness among the users on the one hand and the purchasers of the power driven equipment and machinery on the other. Both will need incentives which shall not be at cross purposes and need not be given in the form of subsidies but basically in the form of tax credits so that they are not constrained by the limited amounts. It should also be remembered that the medium and small scale industries account for 60 per cent of India's exports and any improvement in this sector can significantly contribute in making India's exports far more cost effective.

As far as major industries are concerned, it is necessary to launch immediate campaign in selected industries that are 'large energy consuming' and have the potential to improve through managerial improvements and change in their capital stocks. But this will have a medium term as well as long term perspective.

Transport Sector

As regards transport sector, government's effective Mass Transport Policy will lower energy consumption per capita. But, beyond that, what can be done in immediate term is to convince the individual owners of the vehicles or the transporters on the energy conservation need both from the environmental point of view as well as for making their systems more cost effective. The sector has a very peculiar situation and efforts need to be made on the following two fronts simultaneously:

(i) The manufacturers of the automobiles and transporters must be encouraged through financial incentives and legislative measures to produce energy efficient vehicles.
(ii) The automobile component manufacturers must be encouraged to develop retrofits that will improve the energy efficiency of the existing vehicles.

Certain experiments in the past have shown that energy efficiency upto 30 per cent is achieved if these steps are taken simultaneously. All this can be achieved through awareness creation.

Agriculture Sector

As regards the agriculture sector, efforts would have to be more or less on the lines of the efforts made in the transport sector. Further, it is also necessary to think in terms of operation and maintenance schedules and patterns of maintenance. It has been observed that upto 30 per cent of the fuel consumed can easily be saved if appropriate steps are taken in this sector.

In the agriculture sector better energy conservation is possible through education of the farmers by the dealers of irrigation pump sets. They need to be educated on the higher energy consumption by an inefficient and ill-maintained pump set or equipment vis-a-vis the optimum possible from a similar efficient agricultural equipment.

In addition, rational fuel/electrcity pricing policy shall be another aspect that needs to be properly attended. During the peak seasons, it has been observed that the price of diesel and even kerosene is almost double of the administered price. The user takes it without any murmur. The subsidised price of electricity, diesel oil and other fuels has resulted in unnecessary waste of energy through inefficient equipment and machinery. Appropriate pricing shall be a major contributory factor in controlling consumption.

Power Generation, Transmission and Distribution Sector

Power generation, transmission and distribution losses must be tackled seriously. The government has already identified this sector for private investment, but so far foreign investments are catering to the power generation side only. The investors are not willing to take any risk with regard to collection of money and creating an infrastructure for servicing.

However, for improving efficiency of the power generating system it will be necessary that while selecting the technology for operation the government goes in for the latest technology whether it is thermal power plants, gas based power plants or hydro power plants.

It is also necessary to focus our attention on the production of different sources of commercial energy like coal and oil. This attention has to be paid over the whole life cycle, including exploration, mining, extraction, washing, transportation, refining and ultimate use as a major source and its by-products. A closer study of upstream and downstream production processes and the end use practices will really develop a strategy for overall energy efficiency improvement in this sector.

Future Conservation Strategy

A strategy for energy conservation and efficiency improvement shall have to be developed.

The Approach of the Action Plan for effective energy conservation measure in the country requires following two elements:

(i) *Fiscal Incentives* - the government shall have to provide fiscal incentives linked to energy savings and tax concessions, rather than subsidies;

(ii) *Legislative Measusres* - the government shall have to develop an Energy Conservation Act to enforce punitive action under the law and ensure strict implementation.

A four-pronged strategy needs to be adopted with specific time bound programme. The approach has to be pragmatic. Shorter time period, result-oriented programme should call for lower level of technology interventions and lower financial investments. The four-pronged conservation approach shall be as follows:

(i) Better house keeping which may lead to 5 to 10 per cent of energy saving, on an average, in all the four sectors of use. Such an approach can be followed by inculcating the habit of maintaining systems as per the laid down norms. This shall be a low or no investment approach, but shall be universally applicable/acceptable and affordable in every sector.

(ii) Increased efficiency of existing systems mainly retro-fitting the production units/equipment with more efficient energy use sub-systems which may result in saving, on an average, of 10 per cent and over at a moderate investment level.

(iii) New technologies that call for introduction of state-of-the-art machines, new processes, etc which may help in energy saving of 25 per cent and over, on an average. Such an option is highly cost intensive.

(iv) New sources of energy, mainly renewable in nature, may enable a saving of 50 to 80 per cent. This intervention shall call for change in use pattern and attitudinal change in the user.

A TIME FRAME SHALL HAVE TO BE DEVELOPED

The short-term measures would be in terms of improvements in operational measures and the long-term measures would be in terms of strategic improvements and call for large investments. Thus, the following 3-tier time bound Action Plan is suggested:

Short Term Measures	:	1995-1997 (End of VIII Five Year Plan)
Medium Term Measures	:	Upto 2002(End of IX Five Year Plan)
Long Term Measures	:	By 2007 and Beyond (End of X Five Year Plan)

(i) Short Term Measures

These measures will require intensive efforts to promote the following:

(a) Better housekeeping; and
(b) Increased efficiency of existing systems.

During this short-term phase, initiatives may also be launched to study the options of new technologies and new energy sources.

(ii) Medium Term Measures

These will include continuation of activities launched in the short-term, on which government spending will gradually decrease; and to promote introduction of new technologies in terms of state-of-the-art machines and new processes which, in other words, will be justified because of global industrial competition.

Whereas investments in this phase by the government would decrease, the institutional finance requirement would increase.

The target of energy conservation for the VIII Five Year Plan, set by the government, for 5-year period of 1992-97 has been adopted for 1997-2002, to be achieved in 2002 under medium term of the Action Plan.

(iii) Long Term Measures

These need futuristic biological and bio-technological applications and require R&D support.

CONCLUSION

An effort has been made to outline an approach for energy conservation in the country. However, such an effort would require development of appropriate administrative structures and funding mechanism.

Note: Recommendations of the National Workshop on Strategy for Energy Conservation are given at Annexure-III.

ANNEXURE-1

GOVERNMENT INITIATIVES SINCE 1973

1974 Fuel Policy Committee Report

1974 Indian Oil Corporation—National Productivity Council Directorate General of Technical Development. Fuel Oil Conservation Studies in 300 industries.

1976 Petroleum conservation efforts institutionalised by setting up Petroleum Conservation Action Group, later known as Petroleum Conservation Research Association.

1979 Working Group of Energy Policy.

1981 Constitution of Inter Ministerial Working Group (IMWG) on Utilisation and Conservation of Energy.

1982 Setting up of the Department of Non-Conventional Energy Sources (DNES).

1983-88 Setting up of Advisory Board on Energy (ABE).

1984 Department of Power Identified as Nodal Agency for Energy Conservation.

1987-89 Setting up of the Adviser (Energy Conservation) Office.

1989 Establishment of the Energy Management Centre.

ANNEXURE - II

FISCAL MEASURES

1982 PCRA Boiler Modernisation Scheme.
1983 100 per cent depreciation for notified energy conservation equipment.
1985 Customs duty reduction for fuel efficient vehicles.
1988 Customs duty reduced for notified energy conservation equipment device.
1988 Energy audit subsidy scheme, energy conservation equipment finance scheme.

ANNEXURE-III

RECOMMENDATIONS

A 2-Day National Workshop on Energy Conservation was jointly organised by The Rajiv Gandhi Institute for Contemporary Studies (Rajiv Gandhi Foundation) and Thc Institution of Engineers (India) on September 8-9, 1995 in New Delhi. The participants at the Workshop included policy makers, industrialists, scientists, economists and energy conservation experts. Major recommendations evolved at this National Workshop are as follows:

The following recommendations emerged:

1. The fiscal and physical targets in the energy conservation sector laid down in the Eighth Five Year Plan are not likely to be achieved. This was a matter of concern and immediate action may be required at the national level to develop a strategy to initiate appropriate and relevant energy conservation programme in the country.

2. As an initial step, a non-statutory body of professional experts may be set up to identify the areas for energy conservation in the country and also to develop appropriate fiscal incentives and legislative measures that will bring around the required results.

An important area for consideration of the body would be the mechanism of energy audit in large energy consuming industries, whether through government energy auditors or through an independent forum of energy auditors.

3. Subsequently, a statutory autonomous body at the centre would be necessary to decide about incentives and disincentives; and to have necessary administrative and legislative authority.

4. The autonomous body may effectively implement energy conservation programme through state nodal agencies and professional bodies with built-in flexibility of operation.

5. Energy conservation policy must be developed by the government with time bound targets and clearly defined thrust areas.

6. An Action Plan needs to be developed defining short term, medium term and long term measures and the strategy to be followed for attaining identified targets.

4

NEW APPROACHES IN ENERGY AND ENVIRONMENT INTERACTION

INTRODUCTION

"Role of energy as a prime mover in economic and social development is at times overlooked. The reason basically is economic, rather than of policy. Whenever the price of energy is low, there is a general tendency to develop complacency and take this resource for granted. Very few of the historians would have noted that civilizations have grown and have vanished only for availability of energy or for lack of the same.

Looking back on human history, many scholars have concluded that ancient civilization had an impact on the environment and the natural resources. In all cases where the impact exceeded the environment to cope with it, the civilizations died out. Two typical examples are the Sumerian civilization of Mesopotamia which experienced deterioration of the soil brought about by the salt content in the irrigation water; and ancient civilizations of Greece and Rome, where use of wood as fuel and expansion of agricultural land resulted in destruction of forests. When we in the modern world ponder environmental issues, we would do well to begin by giving some thought to these valuable historical precedents." This is a quote from the presentation of Mr. Shoh Nasu, Chairman of the Board, Tokyo Electric Power Company, at the 16th Congress of the World Energy Council held at Tokyo[1].

The European renaissance was brought around by the value system developed in places like Florence; and the industrial revolution was brought around by energy conversion devices like steam engine. The

1. Key address at the 11th Indian Engineering Congress organised by the Institution of Engineers (India) during December 22-24, 1997 at Bangalore.

climax arrived in the 20th century which has seen one of remarkable economic progress in the developed world. Economic development helped developed countries to support measures for clean technologies. However, the developing countries have not got the economic means to streamline their development. There is deterioration in the quality of life in highly populated developing countries. Poverty and Population are taken as the two 'Ps' responsible for the miserable conditions in the developing world. Very often, it is quoted that poverty is the biggest polluter, and therefore, poverty is the biggest destroyer of environment. Various experts have expressed opinions in different parts of the world. In an Expert Committee set up by the Planning Commission of the Government of India in 1993, under the Chairmanship of Dr M.S. Swaminathan, the well known agricultural scientist, the issue of environment-development interrelationship was thoroughly debated. The Committee had made a very specific recommendation for a paradigm shift in the whole development process. The Indian economy was advised to adjust and accommodate itself to the new challenges so that the growth is environmentally benign.

The overall global economic progress of the 20th century is earmarked by the pursuit of material wealth and has taken advantage of an economic system that utilizes modern science and technology. Mass production has made mass consumption imperative. The industrial growth made the luxury goods available for mass consumption and also generated enough wealth amongst industrial nations so that the common man could utilize those goods. That saw the emergence of a consumer society. The important factor of the industrial revolution in America and the Europe has been the emergence of a middle income group based society and structure which required basic facilities to be made available at an affordable price. That could be achieved only by mass produced items. Energy has always been at the centre of all such issues, and, therefore, the effort has been to keep the energy prices as low as possible.

The development of economically advanced society has led to large scale use of energy to power the modern production systems. Countries that have focused on production processes in relation to their levels of energy consumption have often progressed much faster than the energy rich and complascent countries with efficient technology.

GLOBAL ENVIRONMENTAL PROBLEMS

Global environmental problems have over time developed different characteristics. Three basic characteristics have been often debated on the global platforms.

1. First, the environmental problems originate unavoidably from the economic activities and daily needs of the human society.
2. Second, the counter measures are to be devised urgently, otherwise the damage could lead to permanent or severe impairment.
3. Third, a global response to tackle the environmental issues is required and it goes beyond national boundaries.

The three issues have been easily defined, but when one looks at tackling them the solutions are not found easily. It is necessary to deal with each of them a little more for better understanding.

DIMENSION OF ENVIRONMENTAL POLLUTION DUE TO ENERGY USE IN INDIA

According to data presented by Dr O.P. Rao, CSIR, coal accounts for about 40 per cent of the total commercial energy needs of the country. Coal production through increased opencast mining (which is about 70 per cent of total coal production), its transportation and combustion in the power stations and industrial boilers lead to particulate and gaseous pollution. Coal mining is also leading to land degradation by utilising large area of forest land (21342 hectares between 1990-1995).

Petroleum products account for about 50 per cent of the commercial energy requirements. The consumption of petroleum products in vehicles, industries and domestic cooking produces large quantities of gaseous pollutants. Vehicular pollution is of serious concern in urban areas. Petrol driven vehicles are the major source of carbon monoxide emission contributing about 85 per cent while diesel driven vehicles contribute to 90 per cent of NO_X emissions.

Most of the industrial air pollution is in the form of suspended particulate matter. Emissions from utilities and major industries (textiles, chemicals, iron and steel, cement, fertilizer and paper) are already substantial in many regions of the country and may rise considerably over the next two decades.

TABLE 4.1: Pollutants Emission (in mt_2)

Years	SO_x	NO_x	Particulate
1900	1.0	1.0	26
2000	2.0	2.4	64
2010	4.4	5.0	122

Ref: World Resources Handbook 1994-95.

Indian industries also produce a large quantity of solid waste, for example 30 million tonnes of ash from thermal power plants and 35 million tonnes of slag from iron and steel industry per year. Industrial waste water is over 50 per cent of the total water pollutant load and comes from large and small sized facilities especially in the chemical industry and related sectors. It is estimated that about one third water polluting companies also produce hazardous waste to the extent of 300,000 tonnes per year (1984 estimate).[2]

Indian forests can sustainably provide an estimated 41 million cubic metres of firewood per year, but demand is projected to be 240 million cubic metres which is leading to deforestation.[3]

Over the past decade, the number of agricultural pumps has doubled rising from 4.33 million in 1980-81 to 9.1 million in 1991-92. This resulted in lowering of water table by 0.8 metre annually in some areas of Punjab.[4]

STATUS OF ENVIRONMENT FRIENDLY APPROACH

Energy-Environment issues mainly focus on the carbon dioxide emissions. Most experts focus on methods to control carbon dioxide emissions. The methods to curb carbon dioxide emissions mainly concentrate on supply side and are covered under following categories:

- — higher efficiency for energy conversion.
- — shift towards less carbon containing fuel.
- — imposing a carbon dioxide tax.
- — sequestration by reforestation.
- — use of more nuclear energy.

The demand side receives less attention and focuses only on achieving higher efficiencies in end use devices.

Japan is a peculiar case where government and electricity supply industry have worked hand in hand for over 20 years and have controlled carbon emissions. According to Y. Hosoya[5], actions by ESI have been:

— use "best mix" of primary energy;
— improve the efficiency of the whole energy cycle from source to service; and
— use the best available environmental protection technologies.

Future plans to reduce NO_X, SO_X and CO_2 emissions include the following:

— more nuclear power.
— use renewable energy.
— use urban sources of energy and heat pumps.
— improve efficiency of conversion.
— improve end-use efficiency.

India is in a position to take advantage of both the above routes.

FUTURE POLICIES AND ISSUES FOR CREATING A BETTER ENVIRONMENT

Major global consultations have focussed on following four issues to preserve the environment:

i. Energy Conservation.
ii. Energy Pricing Policy.
iii. Creating Optimum Energy Mix and promoting Nuclear Power.
iv. Technology Transfer.

These issues need some detailed analysis:

(i) Energy Conservation

The first thing we must give thought to is taking care of our limited natural resources and using them efficiently. The best and the most appropriate way to solve the problems of the environment and stabilizing the supply and demand of energy worldwide will be to conserve energy and natural resources.

Conservation of our valuable natural resources is dependent on efficient use and recycling. As economic growth brings material prosperity, we must rekindle in people's hearts the desire to recycle. Such an exercise is already undertaken in a few countries.

Globally, under our present system, it is easy and cheap for individual companies and consumers to engage in mass production and consumption and throw things away. It is unprofitable to conserve energy. The producer companies do so for economy of scales and the consumers do so for reasons of convenience. We must remain aware of the fact that, in the end, the companies and individuals that make up society have a responsibility to the society to preserve the environment. It is the duty of producer companies to limit and soften the effects of their production activities on the environment. Rather than relying on regulation, the companies should devise their own environmental policies. Some major and alert companies in India have in-house groups to focus attention on such issues. But, in ultimate implementation they do not practice unless there is a regulation and strict adherence is called for. A glowing example is the polluting industries of Delhi that have now been closed down and are in the process of shifting outside Delhi.

As for the impact on the environment of the energy consumption required to maintain our lifestyles, it will be crucial for each individual consumer to be aware of the importance of recycling natural resources, and to participate in recycling as a matter of course. Energy Management Centre of the Government of India has promoted some steps but the effect is still not felt to a large extent. Higher level of awareness of the problem alone is not enough. What we need is a system that incorporates a mechanism that ties awareness to actual behaviour.

The problems of energy and environment are intertwined and depend on our lifestyles and consumption pattern. It will be important to implement specific means of energy conservation, in areas such as home appliances, air conditioners, and automobiles, as well as improvements in production efficiencies at the individual company level. At the same time, though, from the vantage point of society in general, we will have to build a system capable of replacing the cycle of mass production, mass consumption, and mass waste. Under present transitory economic conditions in India, appropriate policies are desirable so that market oriented development does not create problems after 10 years, but it should become environment friendly. What we need in India is the modernisation with appropriate adaptation to suit site specific conditions.

In India, the need for energy conservation was felt since early 80s. The Eighth Five Year Plan (1992-97) has appropriately focussed on energy conservation measures and allocated Rs 1,000 crores. However, till the end of the fourth year (1995-96) of the Ninth Five

Year Plan (1995-96) only Rs 50 crore were actually released. One possible reason is that during the Eighth Plan, major challenging target in power generation, coal production and crude oil production were laid down, making experts to believe that additional capacities will fulfill all the demand. The industry has shown reduction in energy intensities. In such places where plant and machinery was totally changed the compound effect of new equipment and conservation measures has been felt. In other cases, good in-house management practices have also resulted in energy conservation—though on a low scale. The important and significant point is the participation of industry and an overall improvement due to their effort.

Energy conservation has to be pursued as a system, through such means and upgrading of our transportation system, recycling consumer goods, and employing unused sources of energy, including waste heat generated by the cities, both in the industrial and domestic sectors. That is expected to take a long time and many be accompanied by some pain.

Such an approach will certainly call for evolving new value system that will be in harmony with nature.

Energy Pricing Policy

The next important issue is the energy pricing policy. Since energy is essential to people's lives and national economies, in some countries people are being provided with energy at below the cost of production for reasons of policy. Yet, it is imperative that energy prices be left to market forces. When energy prices are held down for policy reasons, its consumption will be unnecessarily high. That will impact on the environment and distort the distribution of natural resources. It will also mean the cost of investing in environmental protection cannot be recovered.

In Japan, energy conservation has made rapid strides since the first and second oil crises occurred. In the 20 years between 1973 and 1993, the energy consumption in relation to GDP decreased by more than 30 per cent. Looking at industrial energy, the amount required to produce a tonne of crude steel dropped by 20 per cent. Even in the home electrical appliances such as television sets, air conditioners, and refrigerators became 40 to 60 per cent more efficient during those twenty years.

This progress in energy conservation was occasioned by the rise in energy prices triggered partly by the two oil crises. Yet, in recent years, improvements in energy efficiency seem to have peaked. One of the reasons for that is limitations on technology. But, the main factor seems to be that energy prices have become relatively cheap—almost 30 per cent lower in real terms than in the 70s.

Whereas the hard currency economies have experienced easing of energy pricing burden, the inflation hit developing countries, subject to fluctuating dollar rate parity, are experiencing increasing pricing burden in the local currencies. India is one of them.

The developed countries are moving to introduce and expand competition in the energy market, and to provide people with energy at prices that are as low as possible. In spite of best of the efforts the energy prices in developing countries are going up. More so when the foreign companies enter their markets.

Under the current market mechanism, the cost of dealing with global environmental problems tends to be a diseconomy that is external to the system. The new market mechanism will have to internalise that cost into the mechanism of the economic and social system itself.

As energy pricing and environment preservation pricing are under the umbrellas of two different arms of governments, the environmental costs are never internalised.

The present scenario of making energy available at least price in the developed countries is actually resulting in environment disequilibrium and global environmental pollution.

The situation calls for a new market to trade at prices that fully internalize the costs.

Creating Optimum Energy Mix and Promoting Use of Nuclear Power

The third point is that in order to promote the effective use of limited resources and sustainable development, energy providers must find the "best mix" on both a global and inter-generational basis.

There are various types of primary energy resources, but the key will be to make use of a mixture of energy sources—such as fossil fuels, renewable energies, thorium and uranium. While deciding on energy-mix it will be important to keep in mind considerations of security, economics, and the environment, in order to achieve not only the optimum mix for each nation or region, but also for the world as a whole.

While deciding on the optimum energy-mix it is necessary to take into consideration the level of technology available in the country. India has the option to choose renewables and nuclear both. In case of renewables the technology for large scale operational systems is still establishing; whereas in case of nuclear the technology exists. Then why not pursue this course?

Various experts at the 16th Congress of WEC in October 1995 at Tokyo called for larger base load being shared by the nuclear power companies. Thus a shift from thermal to nuclear plants is a strong possibility. It is recommended that between 1990 and 2020 the nuclear capacity worldwide would not increase from 290 GW to 510 GW, but to 710 GW. If that happens then the carbon dioxide emissions would be reduced from 8.3 Gt to 5.7 Gt which is only 0.5 Gt above the 1990 level.[6] In order to achieve this reduction at competitive electricity prices, societies must change their focus from short term, competition driven investments towards long term commitment. Reliance on market forces alone will not be enough. It has been shown that with standardisation and limitation of the number of reactor models high levels of safety and economical production can be realised. When all external costs would be taken into account, nuclear power would prove to be beneficial for many countries.

At a a Nuclear Power Industry Meet in India held in New Delhi during October 14-15, 1996 the participants echoed similar views. The only difference was that they wanted components to be standardised but not the reactor designs. Also it was suggested that long term plans of 15-20 years need be prepared.[7]

If this can be a reality then use of nuclear power can substantially reduce carbon dioxide emissions without affecting the quality of life.

We need to focus that wisdom on the production of safer nuclear power. Moreover, looking at the nuclear fuel cycle as a whole, we must continue our efforts to elicit people's understanding and trust regarding the comprehensive utilization of nuclear energy.

Technology Transfer

The fourth topic is the need for new international cooperation on energy and the environment. Not only must the various nations and regions of the world come up with their own solutions to problems of the Earth's environment, they must work together as well.

The developed countries, possessing sophisticated environmental technology, must be willing to actively transfer that technology to the developing countries. Unless the package of technologies at a low cost are available to them the developing countries may be forced to opt for the development process that affects environment. Participants to Asia Energy Vision 2020, in New Delhi, in November 1996 also called on the developed countries to fulfil their commitment.

The transfer of environmental technology must not be left entirely to the governmental cooperation alone. Private companies also have a part to play. In their building of energy infrastructure, the developing countries are aggressively adopting technology and funding from corporations in the industrialised nations. Accordingly, the companies in the developed world will make business out of transferring technology for energy utilization to the developing countries.

At the same time, the developed world should consider providing financial support to the developing nations, which because of financial limitations, cannot implement sufficient measures to preserve their environments.

The developing countries should learn from the experience of developed countries and avoid the mistakes committed by them in their development process. The risk is higher at this stage when a number of developing countries are going in for market oriented economies and entering collaborations for rapid industrialisation. Especially important will be the integration of energy and environmental policies with their economic and social planning in the early stages of economic and social development.

The concept of "joint implementation" is contained in the Framework Convention on Climate Change. A trial period that is designed to help establish specific international rules is about to begin. Such experiments are effective means of promoting technical transfer and financial cooperation on the part of the developed countries. I hope that those activities will play a large part in enabling the developed and the developing nations to work effectively together on environmental problems.

CONCLUSION

Energy environment interrelation in India will depend, to a large extent, on the investments provided by the power industry and the policies framed and enacted by the government. Larger foreign and private investment in the power sector will curtail government's intervention purely on environmental considerations. A compromise will be struck to suit the conditions laid down by funding agencies and the technology transferring organizations. Whether India receives the latest and most efficient and environment friendly technology or not will also depend on the capacity to fund projects and programmes.

REFERENCES

1. Keynote Address on Energy for Better Environment (KA), Shoh Nasu, 16th Congress of WEC, Tokyo, 1995.
2. Energy - Environment Interaction - O.P.Rao -IMC-WEC-ASIA ENERGY VISION 2020 - 1996.
3. Biomass Energy Programme for India - David Hall - Dr. Atma Ram Memorial Lecture - 1996.
4. Energy Conservation in Agriculture - M. Sayam National Workshop on Energy Conservation -1995.
5. Proceedings of the 16th Congress of WEC - Tokyo-1995.
6. Proceedings of the 16th Congress of WEC - Tokyo-1995.
7. Proceedings of the Nuclear Power Industry Meet - 1996

5

INDIA'S POWER SCENE AND THE NEED FOR TOTAL QUALITY CONTROL

INTRODUCTION

Energy development in India has been moving at a slow pace during 1994-96 mainly due to financial constraints and low indigenous primary energy resource availability. Coal has remained the major source of energy supply, though the consumption of oil has also been increasing at a higher rate than originally expected.

Whereas resource crunch has adversely affected additional generation capacity, it has also resulted in financially capable industries to think in terms of improving their efficiency of use and extracting maximum useful energy from the primary energy source. Major initiatives introduced by the industry has resulted in 25 major industrial houses receiving the Energy Conservation Award for the Year 1996 for savings in their energy bill up to Rupees three hundred crore per year.[1]

The Ninth Five Year Plan Draft Approach Paper approved by the Union Cabinet on December 23, 1996 has agreed on 7 per cent growth rate for the economy for the period 1997-2002, with agriculture growing at 3.5 per cent and value addition in the industry at 9.3 per cent.[2] This will also enable a 5% annual increase in per capita income. The Planning Commission has estimated that budgetary support required for the Plan will be about Rupees 370,000 crores. The corresponding increase in power sector alone is about 50,000 MW.

PRIMARY ENERGY ENDOWMENTS

Following primary energy recoverable reserves of different sources of commercial energy in India were identified in 1993.[3]

(i) Coal 68,047 mt
(ii) Crude Oil 757 mt
(iii) Natural Gas 686 bcum

LONG TERM ENERGY PLANNING

It has been recognised that the long term energy strategy should be consistent with provision of high priority to meeting fully the basic energy needs of the rural and the urban poor in the immediate future. It should also ensure gradual shift from non-renewable resources to renewable ones with increasing emphasis on demand management, conservation and efficiency.

ENERGY PRODUCTION

Changing pattern of energy supply in India can be realised from the fact that total energy supplies, including both commercial and non-commercial forms, increased from 85.1 mtoe in 1953-54 to about 240 mtoe in 1990-91. The commercial energy contributed 122 mtoe and non-commercial energy 118 mtoe. Fuelwood accounted for 65 per cent of the total non-commercial energy consumed in the country.

Amongst the indigenously produced primary commercial fuels, the relative share of oil and gas has increased from 1.2 per cent in 1950-51 to 32 per cent in 1990-91; whereas the share of coal has declined from 98 per cent in 1950-51 to 63 per cent in 1990-91.

FUTURE ENERGY DEMAND PROJECTION

The projection for future demand for commercial energy in India is as given below in Table 5.1:

TABLE 5.1: Future Demand for Commercial Energy

Fuel Source	1991-92	1996-97	2009-10	2020
Electricity Demand (TwH)	231	336	725	1300 (1745)*
Coal (Million Tonnes)	229	311	650	1 345
Petroleum Products (Million Tonnes)	57	81.2	165	335
Natural Gas (Billion cubic metres)	18.6	30.2	65	135

Ref: Planning Commission, Government of India—September 1995 & September 1996.

* Installed capacity required.

INVESTMENT ON ENERGY SUPPLY SECTOR

The development of commercial forms of energy is highly capital intensive and large investments are required for meeting the rapidly growing demands for these forms of energy from different consuming sectors. The relative share of expenditure on energy sector in the total Plan expenditure during successive Five Year Plans is as shown in Table 5.2.

TABLE 5.2: Plan Expenditure and the Share of Energy Sector

Plan		Plan Expand	Share of Energy Sector (%)				
	Period	(Rs.Cr.)	Power	Oil/Gas	Coal	NRSE	Total
First	1951-56	1960	-	-	-	-	19.7
Second	1956-61	4670	9.7	0.8	1.9	-	12.4
Third	1961-66	8580	14.6	2.6	1.3	-	18.5
Fourth	1969-74	15780	18.6	1.9	0.7	-	21.2
Fifth	1974-79	39430	18.7	3.6	2.9	-	25.2
Sixth	1980-85	109290	19.7	7.8	3.5	0.1	28.1
Seventh	1985-90	220220	17.4	7.3	3.2	0.3	28.2
Eighth	1992-97	434100	18.4	5.5	2.4	0.2	26.5

Ref : Economic Survey Government of India.

Almost 26 - 28 per cent of the plan allocations (in different Plans) have gone to energy sector, and two thirds of this allocation has been for the power sector. As power sector has attracted 18-20 per cent of planned budgetary allocations, it is appropriate to focus on this sector.

POWER SECTOR DEVELOPMENT

Power sector has been very much in the news and, therefore, it is relevant to focus the attention in this presentation only on the power sector. At the time of Independence in 1947, country's total installed generating capacity was about 1300 MW, catering to the electricity demand in a few major urban centres. The present installed capacity of power plants in the country is 83,288 MW (as on March, 1996) comprising 60,087 MW thermal, 2,097 MW hydel and 2,225 MW nuclear.[4]

Present 'All India' maximum peak demand is of the order of 60,000 MW out of which only 49,000 MW could be met resulting in peaking shortage of about 19 per cent. Corresponding energy shortage was around 10 per cent. The current year, 1996-97, is the terminal year of the 8th Five Year Plan and it is anticipated that the All India installed capacity would be raised to about 86,225 MW. The likely achievement in terms of new capacity addition during the Eighth Plan is 18,000 MW as against the target of 30,000 MW.[4]

CEA Study on Demand Projection

The Central Electricity Authority undertook a study in 1991 to project a perspective of national power development up to the year 2006-07 (end of Tenth Five Year Plan). The electricity demand projections as contained in the 14th Power Survey Committee Report form the basis of working out the generation capacity expansion plan up to the year 2006-07 A.D. These studies indicated that to be able to meet peak demand of 144,000 MW and energy requirement of 824 billion units by the year 2006-07, new capacity additions of the order of 142,000 MW, alongwith matching transmission and distribution networks, would be required in the time-frame of 15 years. (The Planning Commission had projected lower additional targets.) The first five years are likely to add only 18,000 MW and it is necessary to add an installed capacity of 125,000 MW to support the planned 7 per cent GDP Growth rate (during the Ninth Five Year Plan and sustain the same in the Tenth Five Year Plan).[4]

This will mean 10-12,000 MW additional capacity has to be created every year. The corresponding investments in capacity addition will be Rs 50-60,000 crores during the Ninth Plan period. Corresponding matching investments in Transmission and Distribution systems will also be required.

Main (and specific) recommendations of the CEA Study for the 'supply' and 'demand' sides were as follows:

CEA Recommendations to Fulfil the Demand

Supply Side

- Double the present pace (year 1990-91) of hydro-power development to add 52,000 MW of hydel capacity.
- Accelerate nuclear power development to create atleast 10,000 MW.
- Reduce T&D losses to atleast 18 per cent from the present level (year 1990-91) of 23 per cent.
- Continue Renovation and Modernisation (R&M) Programme of thermal and hydro units.

— Ensure formulation and operation of National Power Grid by the end of Tenth Plan (Benefit equivalent to 10,000 MW)
— Ensure supply of atleast 15,000 MW from non-conventional sources of energy namely wind, solar, biomas, and micro hydel by the end of Tenth Plan.

Demand Side

— Flatten regional power system load curves to achieve system load factors to above 70 per cent atleast.
— Achieve energy conservation target of at least 10 per cent, 10 per cent and 15 per cent in the domestic, agriculture and industrial sectors respectively.

(The above recommendations may result in capacity benefit of about 30,000 MW).

Identification of Constraints to attract Private Investment

Failure to achieve the Eighth Plan target of 30,000 MW is very often quoted as a major setback to power generation programme leading to slowing down of the economic growth. At the recent International Conference on Energy—Asia Energy Vision 2020 organised by the Indian Member Committee of the World Energy Council, the issue of private investment in the power sector and more effective generation of power were discussed elaborately.[5]

The government explained that the procedures for motivating private investments have to be further streamlined. The health of the SEBs is bad due to non-viable tariff and no-capacity to invest. Financial Institutions have not been coming in a big way. The government has decided to promote independent regulatory authorities at the state level to sanction and promote power plants upto an investment level of Rs 1,000 crores.

The private developers' opinion was that the procedures followed are not realistic and the discussions are not in terms of specifics, i.e., time-frame and details. The independent power producers have to face (i) strict legal requirements of the financial and payment security which is difficult in subsidised tariff regime, (ii) poor PLF and (iii) excessive manpower employed by the utilities.

To top these problems, the primary fuel supplies will also be difficult. The Ninth Plan target for additional generating capacity aims at 50,000 MW needing about 200 million tonnes of coal per year in the terminal year 2007. Against this the coal sector has a plan for an additional mining of 10 million tonnes of coal only per year. Of this, 5 million tonnes will go for meeting the shortages in other sectors and only 5 million tonnes will be available for new capacity. "This would mean a likely capacity addition of about 1,000 MW per year or 5,000 MW only by the end of Ninth Five Year Plan based on indigenous coal as against projected demand of 40 - 50,000 MW", according to Mr. D.V. Kapur, Former Secretary (Energy), Government of India.

New Contracting Procedure for Private Power Producers considered by the Centre

It is reported[6] that the Committee under the Chairmanship of Mr. P. Abraham, Secretary, Ministry of Power, Government of India, set up by the Centre to evolve competitive bidding norms and a tariff structure for private power projects has recommended availability-based bulk tariff, thereby scrapping the present cost-plus approach which guaranteed a minimum 16 per cent return on equity. The Committee has suggested the following approach:

(a) levelised tariff for fixed cost components; and
(b) operational and efficiency parameters such as heat rate and its deration, and auxiliary power consumption.

In the present two-part tariff approach the power producers are tempted to opt for liquid fuel based and gas-based plants as they had an inherent advantage over coal-based plants. The PLF-based approach is being discarded the world over in favour of availability-based approach. The Committee has proposed a cut-off level of 90 per cent availability for liquid fuel-based and gas-based plants and 80 per cent availability for coal-based plants.

This approach will require stricter quality controls.

Suggestions for Future Action

Keeping in view the constraints foreseeable in the energy sector in general, and the power sector in particular, a few concrete conclusions can be drawn:

(i) Formulation of a viable policy for coal production, as coal is going to be the main indigenous fuel for power generation.
(ii) Mobilisation of resources through private power producers and financing agencies to promote independent power

production, as per the internationally acceptable norms and practice.

(iii) Efficiency improvement in the supply side, including reduction in T&D losses, better hydro/thermal mix and development of national power grid.

(iv) Efficiency improvement on the demand side by better energy management and optimal end use consumption.

(v) Energy conservation both on the supply as well as demand side.

Whereas the first two recommendations have emerged out of Asia Energy Vision 2020; the latter three recommendations were mentioned in the CEA Study.

All these tend to indicate that there is a substantial scope for quality indulgence in the power sector from extraction to conversion, transmission and use. An appropriate total quality control by the energy industry and utilities and the total quality management concept in production can go a long way in meeting the energy needs in India by the end of the Ninth Five Year Plan and Beyond.

NEED FOR TOTAL QUALITY CONTROL

The Total Quality Control concept is increasingly applied in the industrial world now-a-days, for ensuring reliable electricity supply and promoting utility management efficiency. The need for introducing this concept in India is felt critically after the Northern Grid Failure on December 20, 1996. Many a contributory factors could have been avoided if only TQC was in vogue in the power sector. Atleast now avoidance of any similar breakdown in future and reliable power supply calls for TQC to be introduced immediately.

Further, the recommendations of Abraham Committee will demand adoption of total quality control by all power producers to honour their contracts.

The quality control has to play an important role both in terms of more efficient utilisation of limited primary fossil reserves in the country as well as in terms of better exploitation of the renewable and nuclear sources.

Whereas in the electric utilities and the energy supply agencies total quality control will result in better quality of energy supply total quality control in the renewable energy sector will help in earlier maturing of technologies that can tap the renewable source at competitive or acceptable costs with an assured degree of reliability.

This will be applicable as much to a modern solar thermal power plant as to a simple solar water heating system, both of them being important for India.

On the energy supply side, the two important aspects (in the Indian context) basically referred to power generation are:

(i) attaining high load factor and plant availability; and
(ii) minimum transmission and distribution losses.

The Total Quality Control concept is essential for these two applications as in some cases the plant load factor in India is observed to be as low as 32 per cent and the transmission and distribution losses are observed to be as high as 27 per cent. Appropriate quality control measures can certainly improve the situation. Through strict quality control measures Kansai Electric Supply Company, Japan, was able to reduce the average service interruption per customer from 7/10th a year to 1/10th a year; and 107 minutes per year down time to as small as 5 minutes per year. At the same time, the Company was also able to ensure safe and reliable power supply through efficient use of men and material resources even from the nuclear power plants.[7]

On the energy application side efforts in the industry have been made to improve the energy efficiency of the products but that is still in the stage of infancy and a major emphasis is required for improving energy efficiency in various applications.

The quality concept has to come from the top and cannot be introduced merely at the bottom. One major reason for positive effect of conservation measures in the industry is the effort of the Energy Management Centre to tackle the issue of quality energy/power utilisation at the level of top management.

Recent example of RAPP II substantiates the view that the quality control measures during replacement of plant components can also contribute to building confidence in the masses about safety of nuclear power plants.

Global cooperation in developing quality control capabilities in the power sector in India is as important as pooling together of the global technological resources for improving power situation in the country. A definite fund allocation for creation of quality control consciousness will certainly help in bringing back returns manifolds compared to the investment. This Fund can help bringing in the best practices from all over the world.

As mentioned earlier, the approach to tap nuclear power, renewables and conservation efforts can result in supply equivalent to

30,000 MW installed, by the year 2006-07. This is equivalent to Rs 120,000 crores investment for capacity addition. Even if 0.1 per cent of this fund is earmarked to TQC, the impact would be immediately felt. Note that this Rs 120 crores does not come from the government alone, but from a number of IPPs and energy industry. If this amount is spread over 10 years, the yearly suggested funding is only Rs 12 crores (which is the cost of 3 MW additional capacity) or just Rs 1 crore per month.

CONCLUSION

The economic development in India needs more energy, mainly electric power, of reliable quality. This can be achieved only through the total quality control in supply and demand sides. Under the new Policy the SEBs, Public and Private Power Producers have to compete with the International Power Producers who have been practicing total quality controls as a work culture. It is time that the utilities, power producers and power industry in India introduced the concept of TQC so that they compete on equal footing with the foreign power producers.

REFERENCES

1. National Energy Conservation Awards 1996 presented by the Ministry of Power on December 14, 1996 in New Delhi.
2. Draft Approach Paper for Ninth Five Year Plan.
3. India's Energy Scenario 1995 - IMC-WEC.
4. India's Power Scene - R.N. Srivastava - IMC-WEC - Asia Energy Vision 2020 - 1996.
5. Proceedings of the 'India Seminar' - Asia Energy Vision 2020, IMC-WEC, 1996.
6. 'Power Projects will not get Guaranteed Returns' - *The Times of India,* New Delhi , page 13, December 26, 1996.

6

NEED FOR THRUST ON R&D IN ENERGY SECTOR

INTRODUCTION

India has passed through difficult energy situation. Recent observations on additional capacity building in different energy sectors hold a grim prospect. The problem is not lack of technology, but the availability of requisite level of investment and latest generation technologies. Whether one blames it to planning or its implementation or the lack of interest of foreign investors and the private industry in supporting the large scale programme in the energy sectors in the country the fact remains that wide gap is now appearing between energy demand and energy supply. A major reason for such a gap is lack of appreciation of R&D.

R&D ISSUES IDENTIFIED BY WEC

In this regard it is relevant to focus attention on the issues in R&D in the energy sector identified at the 16th Congress of the World Energy Council. These can be summed up as follows:

(i) Technical development directed towards improving energy efficiency and competitive supply of renewable energy sources must be accelerated by the reversal of recent declines in research and development expenditure. The effectiveness of Research Development & Demonstration activity will be improved by the development of government/industry partnerships to finance key long-term technologies such as for

---

Paper presented at the Brainstorming Workshop on "Thrust Areas in R&D, Energy Availability and Energy Efficient Technology" organised by DST at IIT, Delhi in New Delhi on September 23, 1996 (paper revised).

new renewable energy sources, and pollution mitigation. Further development of nuclear technology for the full energy cycle also appeared to be needed if nuclear power is to make a major contribution to longer-term needs. Without public confidence in this technology there must be doubts as to the role nuclear energy can play.

(ii) As technology provides greater productivity, and hence additional resources, it should be possible over the next decades to allocate some of these additional resources to meeting the external costs for environmental protection and enhanced efficiency without disturbing the cost to final consumers.

(iii) Over the next 30 years a much wider range of energy forms will be required to satisfy the world's increasing global demand. In addition to fossil fuels, these will include publicly acceptable nuclear development, and those forms of new renewable energy which can be made viable. Governments, in particular, must be encouraged to finance and develop renewable energy to achieve accelerated diffusion, recognising that fossil fuels are likely to remain cost-competitive against alternatives over the next few decades, and are set to play a greater and longer role than was widely understood previously.

The above mentioned observations clearly hint at the major issues in R&D being neglected not only in India but also in other countries.

NEW DIRECTIONS OF R&D

While taking stock of the reasons for the issues identified above, it is relevant to understand the whole process appropriately. Energy supply has two clear and distinct sub-sections.

Primary energy supply is basically an industrial activity which is not having a human interface and normal practices of the industry are applicable to the same. On the other hand, the supply of final form of energy, whether from power station, petrol pump or any other source in electric, liquid, solid or gaseous form, is a service industry and there is a human interface which has its own value judgement. Customer satisfaction is an important element in the second component of the energy systems, whether it is customer driven or not is not the real issue. Whereas the period of 1960—1990 focused attention on the primary form of energy, whether it is coal, oil or nuclear, and tried for means to have additionality of primary resources by identifying higher reserves and the technologies to extract the basic resources from difficult situations, the era from 1990 onwards is now focusing on the final form of energy and therefore; the energy efficiency and the efficient conversion technologies are receiving maximum attention. It really does not mean that energy prospecting of resources has stopped, but it simply means that all energy industries should now focus their attention to serve the customer interest better. And that it where the thrust on the R&D sector has to be.

ENERGY PROSPECTS UPTO 2020

Looking at the thrust areas for energy R&D in India it is necessary to consider the energy perspective in a time frame which, according to the WEC, will be affected by the present day technological know how. Whereas the possibility for projections can be tremendous, it is most appropriate to consider under the business as usual scenario for any projection in 25 to 30 years' time frame. Such an approach assumes that the changes in the consumption patterns of energy, the efficiency improvements, etc continue to take place in future at the same rate at which these were taking place in the past. This logic can always be debated. But, a pragmatic approach based on the business as usual scenario can be the best and most cautious projection to the year 2020. A rough assessment of demand of electricity in year 2020 is 1300 billion units as compared to the actual consumption of approximately 265 billion units in 1993-94. At the present rate of capacity utilisation, the requirement of generating capacity can be worked out to be 378,000 MW. This is roughly four times the present installed generating capacity in India. Demand for petroleum products is provisionally projected at 335 million tonnes in the year 2020. Coal requirement would be about 1,345 million tonnes in the year 2020. Keeping these projections for the demand in 2020, various R&D needs had to be worked out.

PRIORITY IN R&D ACTIVITIES

Whereas the globalisation of economy is making it possible to access the technology internationally, it is becoming really difficult to translate it into practical action. As such, the gap between technology generated elsewhere and technology made available to us is ever widening and not closing. The reason is not basically lack of capacity to assimilate the technology, but the lack of financial capacity to access the modern and latest technology. It may, therefore, be necessary to put a thrust on R&D efforts in these areas to strengthen the indigenous base of technology and to reduce the overall cost through development of appropriate materials, technology and practices. Since electricity is considered to be the most important final form of energy, and therefore, preferred in the planning process, it will play a major role in the substitution of hydro carbons and, therefore, high priority has to be accorded to R&D efforts in both coal and power sectors.

Coal will remain continuously the most important primary source of energy and a new focus has to be placed on coal extraction, transportation and conversion. It is necessary to explore ways and means of coal beneficiation and alternate modes of coal transportation. Indigenous technologies for improving coal quality and improving its end use efficiencies are necessary.

In the oil and gas sector, environmental considerations have been focusing on higher use of gas in the economic and social sectors of the country. Thrust areas in the petroleum and natural gas sector, therefore, will be on finding out various methods for transportation of natural gas, either through pipelines in the natural form or else in the liquid form. At present, as the major cost on gas supply requires capital investment in the transportation sector with a long gestation period, it is extremely necessary to focus attention on R&D in the transportation of natural gas.

WHAT NEEDS TO BE DONE?

A state of complacency appears to have taken over the policy makers in India. The most important object at present is to create an awareness of the need for R&D and mobilising support from the government funding for this kind of R&D. Technological capabilities will neither be transferred from foreign source nor will that be made available free of cost. It will be necessary to provide government fund support for R&D in the country. In this regard, I would also like to refer to a specific recommendation from the Executive Assembly meeting of the World Energy Council held at Kusadasi (Turkey) in October 1996 where its statement said:

> "It is important to revise currently dwindling research and development budgets. The world, by 2050, will require to use all forms of energy to satisfy its requirements, and continued technological advancement is subsequently required for all forms of energy. Particular attention should be paid to the research, development, financing and installation of new renewable (solar, wind, biomass, etc) forms of energy where these are most required,

in the developing countries. With adequate support the share of new renewable energy supplies, currently only 2 per cent could reach 6 - 8 per cent of the substantially increased global energy supply as early as by 2020."

Energy R&D cannot be looked at any closetted sectoral fashion, and only the ministries of Coal, Power, Petroleum & Natural Gas cannot be made responsible for every activity in R&D on their own. It cuts across various disciplines and requires support as a core R&D activity which can only come from the government's major funding under the Department of Science & Technology. The government has to earmark specific budget for R&D in the energy sector and then create linkages with various other ministries and sectoral agencies.

CONCLUSION

Energy needs will continue to grow as India's population will grow; the need for better quality of life and better economic results will be heightened over the next 25 years. Efforts made now will bring in results in over next 10 years or so and implementation of the same can be conducted over 25 years. A well thought out 'policy on energy R&D' has to be the back bone of the 'energy policy' in the country.

Note: The Recommendations of the Brain Storming Workshop on Thrust Areas in R&D for Energy Availability, Energy Conservation and Energy Efficient Technology held on September 23,1996 under the auspices of the Department of Science and Technology, Government of India, have been added as Annexure I to give an idea about the Expert Opinion.

ANNEXURE-I

RECOMMENDATIONS

1. It was unanimously agreed that there is a need to support basic R&D and Demonstration in the energy sector for speedy assimilation of technology and creating matching production and conversion base in the country.
2. The responsibility of organising a Core Group for R&D in energy sector and coordinating with various agencies/ departments should lie with the Department of Science & Technology. No single department/ministry in isolation can look into the R&D problems of the Energy sector, which are multi-disciplinary in nature.
3. The contribution to energy from renewable energy resources in meeting the total energy need has remained marginal. It will be necessary to increase R&D support for advanced technologies/sub-systems based on renewable forms of energy, as well as for the consumer oriented strategies by the industry for demonstration and commercialisation of products developed.
4. Specific thrust areas of energy technology needing R&D support for pulp and paper industry, metal industry, petroleum industry, as well as for electricity, coal transport and fertilizer sectors are as follows:

(i) Harnessing vast hydro potential available which is a clean and efficient energy sources through mini and micro hydel technology with least disturbance of environment and human habitat.

(ii) Development of centralised and stand-alone power generating systems based on renewable energy sources to meet the needs of rural areas.

(iii) Development and adaptation of cleaner technologies, particularly in the field of coal based thermal power generation, with a view to improving the efficiency of existing plants through use of washed coal and introduction of state-of-the-art technology like PFBC and IGCC including retrofit, wherever feasible.

(iv) Development of indigenous capability through adoption of technologies available elsewhere in the world for improvement in quality of power to meet the needs of quality sensitive industries like data management and information technology.

(v) Improvement in energy efficiency in various operations encompassing generation, distribution and utilisation of various forms of energy, with special emphasis on improvements in the existing facilities, better data collection and management and through reliable energy and material audit.

(vi) Focus on transportation sector for improving conversion efficiency as well as reducing environmental pollution by bringing about improvement in indigenous technology.

(vii) Specific studies are required to be taken up on assessment of energy needs of Pulp & Paper industry and rural based energy technologies and assessment of modernisation programme in the power sector.

(viii) Implementation of new technologies like electron beam technologies for producing fertilizers using fuel gas.

5. Finally, there is an urgent need to allocate higher financing provisions for R&D in the energy sector. The government can play the role as a facilitator and promoter. Industry should also come forward to invest in R&D.

7

EXPORT POTENTIAL OF SOLAR THERMAL SYSTEMS

INTRODUCTION

Recent World Renewable Energy Congress at Reading, U.K., in September, 1990, and the Global Conference on Energy & Environment held at Bangkok, Thailand, in November, 1990, projected the possibility of solar thermal systems from India being used in other countries. Its export though at the very preliminary stage, has opened the possibility of Indian manufacturers being able to sell their products abroad not only in the Asian Continent but also in Europe and Africa.

A number of issues regarding quality of products, operation and maintenance and capabilities of these systems have been raised in the past. As the use pattern improved in the country so has the quality and reliability. This being a technology that had to compete with the technology in the conventional energy sector, the economic non-viability has controlled its growth.

Solar energy utilisation in India has been growing systematically and a wide variety of equipment either for heating water or for drying are in use. It can safely be said that they have attained the stage of user friendly status.

IMPACT OF SOLAR THERMAL DEVICES

In places where use of solar water heaters is appropriate for domestic or industrial applications their dissemination will depend on the cost of alternative fuels like oil or electricity. The economics of these devices depends on relative prices or other energy sources.

National Workshop on "Solar Thermal Devices -Commercialisa-tion & Export "organised by the Indian Association for the Advancement of Science on February 14-15, 1991 at New Delhi.

During evaluation of systems installed in Solomon Islands and Vanuatu, it was observed that the large number of solar water heaters in use were promoted due to high cost of petroleum and diesel generated electricity. Neither of these island countries have their large port and as such have to depend on costly transportation systems. Electricity costs are as high as Rs. 4 per unit (1989). The economics of solar water heating systems works very well at these places. Actual payback period for the system works out to less than a year.

It was also observed that the agricultural production in these islands is seriously affected due to poor quality of seeds. Therefore, properly dried seeds can change the shape of things considerably. Such a demonstration in Solomon Islands has already opened the possibility of a large number of seed driers being opted by the users.

Another possibility of solar systems with ready markets is the solar fish drying systems. Most of the developing countries have a large coastline and stable subsistence income generation through artisanal fishery. As the small fishermen do not have preservation possibilities, they catch small quantities of fish and preserve them only by smoking—not a very desirable way but the only available means. All such fishermen are keen to have an alternative cheap method for preservation. Potential use of such systems is high especially in the small island countries or the countries having a large number of islands as these isolated islands do not have any other means of transport and wish to utilise locally available resources most efficiently.

The impact of these devices has been tremendous in all recipient countries. None of these countries have either the materials or the fabrication facilities; the technical manpower for the activities is totally non-existent. They are not in a position to manufacture their requirement. Only option for them is to import from wherever possible.

REASONS FOR MARKET AVAILABILITY FOR INDIA

India has attained a level in production technologies that cannot be matched by others. The potential markets lie in the following regions.

— Pacific Region Small Islands Countries.
— South and South East Asian Countries.
— African Countries.

Approach for different regions can be detailed as follows:

As regards exports to *Pacific Islands,* Australia has the competitive edge. The only country that provides competition to Indian manufacturers is Australia. As the production levels maintained are substantially high and the raw materials are available, they can produce quality products at a reasonably low price. On the other hand they also have the Pacific market at their doorstep. Indian manufacturers in this case may not be able to compete by transporting their equipment from India. However, they have the possibility of Joint Ventures in these Island countries. Shop floor type of production practice is followed by Indian manufacturers on a smaller scale than the Australian or any other manufacturer. This will however, call for the entrepreneurship amongst the Indian industrialists. They will need the support and fiscal incentives from the government under the export scheme.

Next potential market for export of solar thermal products is *Asia.* The market is equidistant from India and Australia. The technological fabrication competence available in these countries is quite high. There is always a possibility of their starting manufacturing to meet their demand. However, assured after-sales service by the Indian manufacturers can fetch them some market.

Solar thermal market in *South Asia* can be captured by the Indian manufacturers. Care has to be taken that the number of individual buyers or the institutional buyers will be low. That reduces the marketing problems. Most of such products will be bought through funds from international agencies like The World Bank, UNDP, USAID etc. The industrialists have to group together with consultancy firms based in India and abroad to activate markets. The industrialists cannot sell products to the funding agencies. It is only the Professional Consultancy Groups that can formulate projects and Schemes to fit into the ideology of funding agencies. Also, the Consultancy Groups only can promote new ideas and related new products. To capture such markets the industrial groups will have to either promote Consultancy Groups or cooperate with the Consultancy Groups.

Africa is the other region that provides ample opportunity for export of Indian solar products. Three basic elements of situation that exist in Africa are: cost of electricity for domestic and industrial consumption is high, labour is costly and the cost of materials is high. All these make the possibility of production of solar equipment at a reasonable cost a remote possibility. Under these circumstances the Indian manufacturers can supply products to Africa at a low cost and caputre market. At the same time, the geographic situation does not permit Austrialian manufacture to be cheaper. Another major advantage in Africa is that a large number of donor agencies are operating and are willing to provide support. Even such international funding agencies like USAID that normally provide for professional services also provide hardware costs for supply to Africa. Africa Fund is another major funding agency that provides funds for equipment. Incidentally under this programme every country can only be providing hardware and not cash grants. Therefore, it becomes much easier for the Indian manufacturers as they have only to approach the Indian government provide funds or include supply of solar equipment under this scheme. This is, however, only one part of the story. The demand for such equipment has to emerge or eminate from the receptor country. To fulfil that requirement it is necessary for the manufacturers to liaise with the active groups in Africa or consultancy firms who could provide with the actual requirement. Otherwise, a survey will have to be conducted reflecting demand from Africa.

Equipment to be exported will fall under two categories.

(i) Totally manufactured items like solar water heating systems which will only have to be installed. (A do-it-yourself instruction book can do the work.

(ii) Partly manufactured items and partly fabricated and assembled, with civil works being done at site like solar driers.

The manufacturers do not need skilled personnel for the first category of contracts, but they do need skilled manpower for the second category for execution of projects. This skill will basically be of the technician grade for installation and of higher level for part manufacturing and designing in India.

Therefore, proper assessment of requirement shall have to be undertaken. Maintenance of such staff will be a costly affair.

CONCLUSION

This potential cannot be tapped unless government provides appropriate package of fiscal incentives. This package shall have to be worked out mutually by the manufacturers and the government.

Note: Recommendations of the National Workshop on "Solar Thermal Devices—Commercialisation & Export" are given at Annexure-I and the Plan of Action for Export is given at Annexure-II.

ANNEXURE-I

RECOMMENDATIONS

Thrust on Rural Application and Additional Subsidy for the Same

A number of solar thermal systems are ideally suited to preservation of agricultural produce and for food processing industry. Wider utilization of solar thermal decies should be promoted in these sectors.

Ministry of Agriculture and Co-operation should be approached for additional subsidy for rural and agricultural applications as is practiced for tractors, pump sets and agricultural inputs.

Incentives for Technology Upgradation

Incentive and support for upgradation of product technologies and production technologies should be provided to the industry.

Support for inhouse research and development should be provided and the ownership rights of developed technology should be left with the industry.

Tax & Duty Exemption on Components

Tax and excise duty on components should also be exempted as is available for the total solar product. This will help in development of specific component industries and the high level of production will help in the introduction of mass production techniques.

Subsidy to Users

Subsidy and tax concessions, by way of 100% depreciation allowance. both should be provided on domestic solar thermal systems as in the case with the industrial solar thermal systems.

Terms of IREDA Funding

Procedures for IREDA Funding should be made easier and the initial fee for registration should be rationalised on percentage basis subject to a maximum of Rs. 3,000 rather than a flat rate of Rs. 3,0000, which is too high for small system.

Thrust Areas in R&D

Major areas of R & D activities for future commercialisation of technologies need to be identified and R&D in industrial establishments in addition to the academic institutions need be promoted.

Awareness Campaign

It is necessary that major awareness campaign is launched in the mass media, both electronic and print. To ensure the returns, quality film programmes and write ups should be supported even from private producers. The campaign has to be aggressive in nature to carry the message.

Training Programmes

Support for manpower development by way of supporting regular training workshops and courses for different levels of technical skills is necessary.

Specific training programmes for manpower development at the technician grade for installation, insulation and maintenance need to be undertaken.

Safeguarding User Interest

To safeguard users interest supply of Operation Manual along with the solar thermal system should be made mandatory.

Subsidy Disbursement

Uniform subsidy disbursement policy should be evolved so that delayed decisions do not act as brakes in the increasing utilisation of solar thermal devices.

Extension of Solar Passive Systems

Demonstration/Extension programme may be adopted for promoting passive applications.

ANNEXURE-II

EXPORT OF SOLAR THERMAL DEVICES PLAN OF ACTION

Strategy

The export of solar thermal equipment could be planned on the following lines:

(i) Adequate exposure of the capabilities and competence of Indian units to overseas buyers, especially in potential markets.
(ii) Adequate exposure to Indian manufacturers of international market and state of the art technology.
(iii) Organisation of suitable forums for interaction between Indian exporters and overseas importers:
(iv) Creation of appropriate mechanism for regular dissemination of market information of Indian exporters.
(v) Continuous dialogue with the Indian exporters on issues connected with export and steps to remove bottlenects in export.

Potential Regions for Export

Potential regions for export of solar thermal devices are as follows.

(i) South East Asia
(ii) South Asia
(iii) Africa

Pacific Island countries also can be tapped out mainly for the joint ventures.

Modes of Export

Solar thermal equipment can be exported in the following three different modes:

(i) CKD packs—for products like solar water heating systems, solar cookers and solar distillation system.
(ii) Partly finished packs—for products like solar driers.
(iii) Joint ventures for manufacture of products in the other countries

with the Indian technical know-how and components.

Levels and Forms of Technological Requirement

Following levels of technological skills for export will be required:

(i) High production capability, for CKD packs, in the country.
(ii) High technical skills, for fabrication of equipment, in the country and high technical skills for technican grade staff for installation and maintenance in the purchasing countries;
(iii) High technical capability, for transfer of production technologies and operation of joint venture, in the recipient countries.

Funding

The exports to South East Asian Countries will be mostly on purchase of individuals and institutions.

Export to South Asian and African countries will be mainly on the funding provides by international funding agencies like World Bank. UNDP, USAID, SIDA and Africa Fund.

Project preparation for funding by international agencies will require involvement of consultancy organisations operating at National and International levels.

Export Incentives

The package of incentives available to Engineering Industries may also be adapted for industries in Export of Solar products.

Market Survey

Market Survey can be conducted through the Engineering Export promotion council.

Marketing Strategy

The marketing strategy will require creating a good image of Indian goods and Indian industrialists. Engineering Export Promotion Council can play an important role and their office can be used for organizing the following:

(i) Buyer-Seller Meet:
(ii) Exposure of the Indian Industrialists to outside markets:
(iii) Image building of the Indian industry abroad;

To do so, Engineering Export Promotion Council will require quality publicity material on government's approach and industrial products in the print form and the audio visual form.

8

MAKING SOLAR WATER HEATERS WORK

INTRODUCTION

Solar Energy has been traditionally used for low temperature in the country since ages. Various efforts at improving the output and the productivity of solar thermal devices led to involvement of major research institutions in the country into development of such devices. National Physical Laboratory in New Delhi and the Central Arid Zone Research Institute in Jodhpur have been two pioneering institutions focusing attention on low temperature solar thermal applications. Incidentally, it was National Physical Laboratory that thought of developing a Perspective Plan for the use of solar energy at the time of the first oil crisis in 1973. That pioneering effort led to documentation of possibilities of solar thermal applications in day-to-day life of the common man.

The Department of Science and Technology launched a major movement for use of renewable source of energy by setting up a Commission for Additional Sources of Energy in 1981. The Department of Non-conventional Energy Sources was created in 1982 which has now been upgraded to a Ministry. Consistent efforts of the last 15 years have helped to improve the level of technologies developed in the country and thereby their acceptability. Intermittent reviews have also helped in improving the effectiveness of these technologies. Planned effort have resulted in awareness creation, technology proving and commercialisation of low temperature solar thermal devices. The impact is now being felt in different parts of the country.

Address at the Seminar on "Solar Water Heater Systems" organised by the C.P.W.D. Training Institute at New Delhi during January 9-10,1997.

ROLE OF INDUSTRY

Involvement of the industry in this sector was emphasised right from the very beginning. As large markets did not exist, it was always a small scale or a medium scale enterprises that entered the solar thermal market barring exceptions. Uncertainty of market and lack of skills in formative years of state government nodal agencies led to slow growth of the solar industry. It is often said that the opportunists entered the market and vanished. This may not be very true. Every industrialist is supposed to enter the area where there is an opportunity. Total uncertainty could attract only the enterprising and those willing to take more risks. However, their capacity to sustain the market risk was also low and as such they could not sustain losses for a long time. There was a need for developing a market to a sustainability stage but that phase was missing and the solar thermal market moved from subsidy based administered prices to market based pricing system.

The present day industrialists are not dictated by the designs developed and approved by the state nodal agencies, thereby forcing them to tell the customer that what he is selling is really beautiful. It is now the customer's choice and only when he considers the product beautiful is that accepted. The competitive spirit is now entering into the current entrepreneurs.

Whereas the Bureau of Indian Standards has laid down certain performance standards and manufacturing standards, what is still lacking is the standard designs. This standardisation is not necessary for the system as a whole, otherwise the uniqueness of the product design for different manufacturers will vanish. That may even deprive them of their competitive edge. What is necessary under standardisation is the standardisation of the components.

The ultimate in market orientation to solar thermal products will come only when the specifications are properly defined by the manufacturers and the defaulters are subject to penalties. The situation is gradually changing and accountability of manufacturers is increasing.

CUSTOMER PROTECTION

The customer is also interested in getting a reliable product that will provide him the kind of promised service for the time period stipulated under the purchase agreement. The Consumer Court provides the protection to the customer for the same. The Bureau of Indian Standards has laid down norms for manufacture of equipment for solar thermal devices. Standardisation on one hand and the Consumer

Court Protection on the other is now the 'safety-net' for the customer.

Large scale purchases, like of CPWD, can build-in a 3 year or 5 year maintenance contract in the installation tender to ensure trouble-free running for 5 years. The manufacturer will be the best person to maintain such system. CPWD will be benefitted by better accountability of the manufacturer.

CASE STUDY

The Indian Association for the Advancement of Science, under my guidance, had carried out a Case Study of operative solar water heating system and conducted their cost benefit analysis, in 1987. (The findings were also presented to the Solar Energy Convention in December, 1987 and refer to prices prevailing in 1987.)

The Study covered following categories of end-users:

(i) Textile industries
(ii) Dairies
(iii) Breweries
(iv) Hotels
(v) Educational Insitutions
(vi) Hospitals.

The size of solar water heating system covered under the study varied from 4,000 litres to 60,000 litres.

The selection of the systems was so made that different categories of applications were fully covered and the geographic locations were as wide and disperse as possible. Also these systems were replacing the existing devices that worked on different types of fuels: coal, fuel wood, electricity and fuel oil. Data for their operation for atleast two years was available in every case. In some cases, it was available for five years. Based on that, various calculations were made to reflect the savings in different parts of the country. Subsequent calculations were made for establishing the cost benefit of such systems.

Enclosed Tables give details of observations. With a view to give a better understanding, the Tables have been separately discussed.

Table 8.1 indicates that the cost of 1,000 KCal useful heat from coal was available at 35 paise where as that from electricity was at Rs.2 or almost three times the cost. While working out these figures, efficiency of use was also taken into consideration. The ratio still holds true although the pricing of the sources of energy have changed.

TABLE 8.1 Cost of Useful 1000 Kcal Heat from Different Sources

Sr. No.	Source of Energy	Calorific Value	Cost of Energy	Efficiency of Use	Useful Calorofic Value Available for Water Heating	Cost of Useful 1000 K Cal
1.	Electricity	860Kcal/KWH	Rs. 0.75***	90%tt	770Kcal/KWH	Rs. 1.00@'
2.	Fuel Oil	10,000 Kcal/Kg 8,500 Kcal/Ltr	Rs. 3.75	70%t	6,000 Kcal/Ltr	Rs. 0.60@
3.	Coal	5,000 Kcal/Kg	Rs. 1.00	60%t	3,000 Kcal/Kg	Rs. 0.35#
4.	Firewood	2,000 Kcal/Kg	Rs. 1.00**	60%t	1,200 Kcal/Kg	Rs. 0.80@"

* Prices in Delhi

** Price of Firewood can vary from Rs. 0.50 to Rs. 1.00, Depending on Site

*** Price of Electricity varies from Rs. 0.75 to Rs. 1.50/KWH

Price of Coal and Fuel Oil is more or less uniform.

t° Efficiency of Conversion in Boiler

tt Efficiency of Conversion in Geyser.

@ Figures Rounded-off.

' Price Varies between Rs. 1.00 and Rs. 2.00/1000 Kcal

" Price Varies between Rs. 0.20 and Rs. 0.80/1000 Kcal.

Table 8.2 reflects the cost of 1,000 LPD system for temperature of operation at 60 degree centigrade and 85 degree centigrade.

TABLE 8.2: Fuel Cost of Heating Water in 1000 LPD Systems

Sr. No.	Source of Energy	Price of 1000 KCal	Fuel Cost of Heating Water	
			60C Systems (40,000 KCal)	85°C Systems (65,000 KCal)
1.	Electricity	Rs. 1.00	Rs. 40.00	Rs. 65.00
2.	Fuel Oil	Rs. 0.60	Rs. 24.00	Rs. 39.00
3.	Coal	Rs. 0.35	Rs. 14.00	Rs. 23.00
4.	Firewood	Rs. 0.80	Rs. 32.00	Rs. 52.00

1. Large Solar Water Heating System at 60°C Heat Input

Rise in Temp = 60-20 = 40°C [20°C Taken as Average] Cold Water Temp

Heat Input for 1000 Ltrs = 40,000 KCal.

2. Large Solar Water Heating System at 85°C-Heat Inpur

Rise in Temp = 85-20 = 65°C

Heat Input for 1000 Ltrs = 65,000 KCal

Table 8.3 is a compilation of details of all the systems studied at that time. The table gives the capacity of the system and the type of fuel being replaced; temperature in degree centigrade; annual utilisation

TABLE 8.3: Large Solar Water Heating Systems Studied

Sr. No.	Site	Capacity LPD	Temp °C	Annual Utilization of Water K.Ltr.	Capital Investment/ (Users Share)	Annual Energy Saving m KCal	Annual Fuel Cost Saving/ Users view	Annual Fuel Saving Units	Infrastructural Cost Saving
	Textile Industries								
1.	Malviya Mills	12,000 C	85	3,600	Rs. 6,00,000	234	Rs. 83,000	83,000 kg	Rs. 64,740
				(Nil)		Rs. 40,000			
2.	Western India Mills	6,000 F	85	1,800	Rs. 4,00,000	117	Rs. 70,000	18,700 ltr	Rs. 37,400
	Dairies			(Nil)		Rs. 55,000			
3.	Lal Kuan	10,000 C	60	3,000	Rs. 8,00,000	120	Rs. 76,500	76,500 kg	Rs. 59,670
		5,000	85	1,500	(Rs. 2,00,000)	87.5	Rs. 2,58,000		
4.	Lucknow	10,000 F	60	3,000	Rs. 6,00,000	120	Rs. 1,18,800		
		4,000	85	1,200	(Rs. 200,000)	78	Rs. 2,55,000	31,680 ltr.	Rs. 63,360
5.	Haryana Breweries	15,000 F	60	4,500	Rs. 6,00,000	180	Rs. 1,00,000		
	Hotels				(Nil)		Rs. 1,00,000	28,800 ltr	Rs. 57,600
6.	Janpath	40,000 F	60	12,000	Rs. 14,20,000	480	Rs. 2,88,000	76,000 ltr	Rs. 1,53,600
					(Nil)		4,00,000		
7.	Lodhi	60,000 F	60	18,000	Rs. 2,80,000		Rs 4,32,000		
	Educational Inst.				(Nil)	720	Rs. 6,00,000	1,15,200 ltr	Rs. 2,30,400
8.	Sr. Dasmese Academy	14,000 W	60	2,100	Rs. 4,42,000	84	Rs. 34,000		
					(Rs. 479,000)		Rs. 35,000	66,000 kg	Rs. 82,500
9.	Delhi Public School	5,000 E	60	750	Rs. 2,00,000	30	Rs. 30,000	40,000 kg	
	Hospitals				(Nil)				
10.	Holy Family	12,000 E	60	1,800	Rs. 5,30,000	72	Rs. 72,000		
					(Nil)		Rs. 45,000	86,000 kg	
11.	A.I.I.M.S.	14,000 E	60	4,200	Rs. 6,85,000	168	Rs. 1,48,000	1,48,000	
					(Nil)		Rs. 1,28,000	2,24,000 kgs.	

W — Firewood Operated C — Coal Operated F — Fuel Oil Operated E — Electricity Operated of water in kilo litres; capital investments and the users share; annual energy saving; annual fuel cost saving - both normative and users reported value; annual fuel saving in physical savings and infrastructure cost saving.

As the savings against the electric devices were expected to give the best figures and the best pay-back time period a study of three solar water heating systems at Holy Family Hospital, AIIMS and DPS in Delhi was conducted to determine the total capacity requirement and the electricity consumption in the normal process and the effective savings by use of solar water heating systems. Cost comparisons were accordingly made. It showed that with a capital cost of Rupees 25 lakhs the national exchequer saved Rs. 2.5 crores. At present costs these figures will be Rs. 50 lakh and Rs. 4 crores. Refer Table 8.4.

TABLE 8.4: Energy Saving & Installed Capacity Saving

(Compared to Electrical Devices)

Calculation Based on Daily Use

Sr. No.	Site	System Size LPD	Energy Needed KCal	Supply load kwh	Supply load with 25% overload	Supply Cap. if Heating is done in 8 Hours
1.	Holy Family	12,000	480,000	623	778	97 KW
2.	A.I.I.M.S.	14,000	560,000	727	908	113 KW
3.	D.P.S.	5,000	200,000	260	325	41 KW

1. 30,000 LPD System Need, 250 KW Load. As 1 KW transmitted needs 2.5 KW installed generating capacity for 250 KW load will need = 250 x 2.5 = 625 KW installed. 50,000 LPD System will need 400 KW load & 1 MW Installed. Capital cost of 50,000 LPD System Rs. 25,00,000 Capital cost saving on avoiding 1 MW installed capacity Rs. 2.5 crores (rs. 2.16 ABE)

2. If auxiliary losses of 5% and transmission & Distribution losses of 21% are considered, only 74% power is transmitted

Installed generation Capacity = 400 X 100/74

For 400 KW Transmitted = 540 KW

Capital Cost of 540 KW Installed = 540X2.5 ABE 540 X 2.16

(without Provision for P.L.F.) = Rs. 1.35 Crores Rs. 1.134 Crores

The Table 8.5 reflects the total savings from all the systems studied under the survey.

For all year round system, the pay, back period varied from 4 years to 11 years, depending on the kind of system and the fuel being replaced. Refer Table 8.6.

The economics and the cost benefit was carried out and suitable sites and excellent sites were identified for different fuel replacement

TABLE 8.5 : Total Energy Saving Possible from Solar Water Heating Systems

Sl.	Description	Total Capacity Sanctioned LPD	Energy Saving in a Year m Kcal
1.	Large	3,416,950	41,003
2.	Domestic	166,700	1,000
	Total	3,583,650	42,003
		= 3,600,00	

Note:

(i) No. of Operating Days for Large Systems—300 per Annum

(ii) No. of Operating Days for Domestic Systems — 150 per Annum

(iii) Energy Saving Calculated for Systems to be operating at 60°C

Actual Saving May be higher as some systems will be operating at 85°C

Total Capacity of Solar Water Heating Systems — 36 Lakh L.P.D.

Total Installed Capacity Reqd. if only electricity is used = 72 MW

Total Saving in Capital Costs — 72 x 2.5 = Rs. 180 Crores

Total Expenditure in Capital Costs @ Rs. 50/L.P.D. = 36,00,000 x 50 = Rs. 18 Crores

Total Saving in operation @ Rs. 12/ltr Per Year — Rs. 4.32 Crores

Total Installed Capacity Reqd. (without consideration for P.L.F.) = 38.88 MW

Total Saving at A.B.E. Costs = 38.88 x 2.16 B83.99 Crores

and giving the best payback period. Refer Table 8.7

Different scenarios were developed to indicate the possibility of government sharing the investment both in terms of providing subsidies as well as in terms of providing soft loans.

TABLE 8.6: Annual fuel Savings & Payback Period Rule of Thumb

Installation Cost — Rs. 50 LPD (60°C)

A) Year round operational system Working days -300 per year.

Sr.	Comparative Energy Source	Savings/ Ltr. Installed	Payback Period
1.	Electricity		
	@ Rs. 0.75/kwh	Rs. 12.00	4 Years
	@ Rs. 1.50/kwh	Rs. 24.00	2 Years
2.	Fuel Oil @ Rs. 3.75/Ltr	Rs. 7.00	7 Years
3.	Coal @ Rs. 1.00/Kg	Rs. 4.50	11 Years
	@ Rs. 1.00/(85° Systems)	Rs. 7.00	7 years
4	Firewood @ Rs. 1.00/Kg	Rs. 5.00	10 Years
	@ Rs. 1.00/Kg (85°C Systems)	Rs. 8.00	6 Years

Savings are Directly in Terms of Savings in Fuel Bill and do not include savings in maintenance of conventional water heating devices.

B) Six month operational systems (Working days - 150 per year)

1.	Electricity		
	@ Rs. 0.75/kwh	Rs. 6.00	8 Years
	@ Rs. 1.50/kwh	Rs. 12.00	4 Years

Six month operational systems are mainly run on electricity. Thus only Electric systems for hostels, educational institutions and residential buildings considered.

TABLE 8.7: Economics & Cost Benefit

Suitable Sites	
Year—Round System	
Fuel Oil Replacement (60°C Systems)	Payback 7-8 Years
Coal/Firewood Replacement (85°C Systems)	in Terms of Fuel Bill saving
Half-Year Systems	
Electricity Replacement (60°C Systems)	
(Cheap Tariff Sites)	

Excellent Sites	
Year-Round Systems	
Electricity Replacement (60° & 85°C Systems)	
(All Tariff Sites)	
Half-Year Systems	Payback 2-4 Years
Electricity Replacement (60 & 85C Systems)	In Terms of Fuel Bill Saving
(High Tariff Sites)	

Table 8.8 gives an overall picture of the infrastructural capital cost of sources of energy; capital cost of end-use device; total capital cost; operation and maintenance cost; fuel cost and total operational cost.

The Case Study was conducted for the first time and helped in authenticating the savings and payback periods.

TABLE 8.8 : Capital Cost & Operational Cost using Different Energy Sources—50,000 LPD System operating at 60°C

Sr. No.	Source of Energy	Infrastructural Capital Cost of Source of Energy '000 Rs.	Capital Cost of End Use Device '000 Rs.	Total Capital Cost '000 Rs.	Operation & Maintenance '000 Rs.	Fuel Cost '000 Rs.	Total Operational Cost '000 Rs.
1.	Electricity	1,134 Crores* 2,500 crores	6,00	1,194 Crores 2,560 Crores	30 1,80 2,10	600	8,10
2.	Oil	1,92	4,30 (2 Units)	6,22	1,50 1,80 3,30	360	6,90
3.	Coal	1,63	8,00 (2 units)	9,63	1,50 1,80 3,30	210	5,40
4.	Firewood	2,10	3,50 (2 Units)	5,60	40 1,80 2,20	4,80	7,00
5.	Solar	Nil	25,00	25,00	36	Nil	36
6.	Solar with Standby	32	25,00 2,35 27,35	27,67	36 25 30 91	60	1,51

Note on Infrastructural Capital Investment Values

1. Coal — Rs. 780/T [ABE As per VII Plan]
2. Firewood Rs. 1,250/T [Dnes Project Cost for Plantation Raising]
3. Electricity Rs. 2.50 [With P.L.F.] Rs. 1.134 [Direct Trans/KW]
4. Oil — Rs. 2000/Kltr.
5. Solar Standby for 60 Days

*Transmission Basis

9

TRANSFER OF SOLAR TECHNOLOGIES IN INDIA: Some Issues

INTRODUCTION

Energy consumption patterns in India indicate that on one hand energy consumption in the urban sector is comparable to any developed country, in the rural sector consumption is only for basic sustenance of cooking. Overall, India's per capita consumption of commercial energy is only one-eighth of the world average and will increase with the growth in GDP and improvement in the standard of living. A major challenge, therefore, is to meet the growing requirement of energy in the rural areas. Rural electrification has successfully carried electric power to a large number of villages that are easily accessible to the national grid. However, other sources of non-commercial forms of energy also play an important role. The principal sources of renewable energy are solar, wind and bio-energy. Decentralised micro-hydel plants have also been recently tapped. Solar energy, however, has resulted in development of an acceptable form of energy.

SETTING UP OF DNES

In view of the suitability of renewable energy sources a high-powered commission namely Commission for Additional Sources of Energy, CASE, was set up in March 1981 for the purpose of developing and promoting New and Renewable Sources of Energy. The responsibility of the Commission was defined as formulating policies and

---

Address at the UNDP sponsored National Workshop on Technology Transfer in Solar Energy Applications organised by the Solar Energy Centre at New Delhi during November 18-22, 1991.

programmes for the development of NRSE, coordinating and intensifying R&D activities in this area and ensuring implementation of the government policies in regard to all matters concerning NRSE. The Commission was required to provide necessary linkages between R&D and production by promoting the acquisition of technical capability and providing finances for development and rapid commercialization of NRSE. The Commission further recommended to the government the incentives for the commercial use of NRSE technologies by industries, advised the government on import of technologies in this area and is responsible for operating all necessary industrial and import licensing policies and procedures in the area of NRSE. Initially, the Commission functioned in the Department of Science and Technology. The Government set up a Department of Non-Conventional Energy Sources on 6th September, 1982. In the Allocation of Business Rules, DNES was assigned CASE as its primary responsibility. The Secretary, DNES, was also appointed as Chairman, CASE.

Considering the initial requirements of this Department, the Scientific Advisory Committee to the Cabinet included DNES in the list of Scientific Departments and Agencies.

Objectives of Solar Thermal Technology Programme

Solar Energy Technology Programmes have found a special place in overall development of new and renewable sources of energy. In its 15-year profile, solar thermal programmes were expected to achieve the following:

(i) utilise solar thermal energy to meet cooking energy requirements and hot water requirements in the domestic sector, particularly in rural and remote areas;
(ii) utilise solar thermal energy for purification of brackish water by solar distillation;
(iii) provide heat requirements for drying agricultural and horticultural products;
(iv) utilise solar energy for refrigeration and cooling needs for preservation of food products, vaccines etc with particular emphasis on remote areas;
(v) utilise solar thermal energy for industrial heat requirements;
(vi) utilise solar thermal energy for generation of power;
(vii) promote and assist research and development activities for the development of devices, materials and systems for most

economic utilisation of solar thermal energy;

(viii) promote industries for production of materials, components and systems for efficient utilisation of solar thermal energy.

Current Policy of DNES

The Current Policy has following elements:

i. *Awareness*

Utilisation in a scientific manner has to be supported by acceptability of people. In this regard, awareness through setting up of domestic plants; setting up nodal agencies in each state; providing financial assistance, incentives for installation of solar thermal systems/devices; and use of mass-media have been considered important aspects.

ii. *Research and Development*

The solar energy activities initially started as R&D activity in major laboratories and educational institutions. As a result of sustained efforts, the production technologies for solar cookers and solar water heating systems have reached a commercial state. Further, R&D in solar refrigeration and air-conditioning, solar power generation, solar thermal pump, solar passive architecture, solar pond and component and material development is still to be supported.

iii. *Fiscal Incentives*

The capital expenditure on conversion and utilisation systems in solar energy application is higher than the capital expenditure on utilisation systems only in the case of conventional sources of energy. To share expenditure and to provide financial support to the manufacturers and to the users the government has three different types of incentives, viz., subsidy, tax concession and loan, as detailed below:

Subsidy is provided as a direct sharing of the cost as a percentage of the price of the system. The concept of subsidy in case of renewable system is totally different from that of subsidy, say, in case of agriculture. In case of renewable system, the subsidy is aimed at not only promoting these devices but is also aimed at conserving the fossil sources of energy.

Tax concessions are provided in the form of exemption of sales tax and excise duty on the manufacturing side and the income tax for the users. Whereas the Central Government has a set of incentives, the State Governments have also provided exemption of sales tax and some other incentives.

Loan from financial institutions is provided for the manufacture, mainly the priority loan under the IDBI's direct refinance schemes. Indian Renewable Energy Development Agency, set up under DNES, also provides soft loan at 8.5 per cent interest. This is for both, the capital expenditure as well as for purchase of the equipment by the users.

iv. *Standardisation*

The Bureau of Indian Standards has laid down a number of standards for conducting test on equipment and components with a view to standardisation of various solar devices in the country.

INITIAL IMPACT

Normally a decision is based on the experience of various systems and devices in the country. There have been extremely successful cases of use of solar systems but there have also been cases where the benefits to the users have not been upto the expectation. In case the system was installed on a commercial organisation, that has been found working regularly and continuously and properly maintained by the user agency, as they have the capability to understand the technology and also appreciate the saving of fossil fuel. On the other hand, in case of systems installed on government buildings as demonstration units, the response was very poor and they were often found shut down for no reason. This clearly demonstrates that in a receptive environment the technology transfer is carried out effectively than in a neutral environment. The hypothesis here is that the caretakers in the government buildings, where such systems were installed as demonstration pieces, were not opposed to the idea but were unconcerned.

In case of the domestic sector, successful demonstrations at exhibitions, fairs etc and individual user demonstration have tremendously helped in creation of awareness of such devices. Depending on the level of manufacturing capabilities and the area of operation of the supplier, domestic systems have worked satisfactorily under certain circumstances and have also failed miserably under different set of conditions. At the same time, non-availability of equipment in the market in the off-the-shelf form killed the enthusiasm of the potential buyers.

Thus, there are successes and failures of the same technology under different use and environment conditions.

DEMAND POTENTIAL IN 2000

Solar assisted hot water systems, solar cookers and solar stills are found useful in the domestic sector on a representative basis. Keeping in view the initial penetration rate and the projections of the other authentic institutions, the Indian Association for the Advancement of Science, under my guidance, has made the following projections:

Solar water heating system potential by the year 2000 can replace fossil energy to the extent of 61,70,400 m.kcal; solar cookers can replace 15,86,000 m.kcal; and solar stills 44,40,631 m.kcal.

The solar assisted hot water and air use in selected industries by year 2000 was also conducted. Following groups of industries were found suitable for such use: (For appropriate levels of penetration rates, different energy quantities were worked out)

Hot Water (Group A)

1. Dairies
2. Hotels
3. Textile Mills
4. Edible Oil
   (a) Extraction
   (b) Refined
5. Breweries
6. Distilleries
7. Bulk Drug Industries

Hot Air (Group B)

1. Tobacco
2. Tea
3. Wood Seasoning

The energy saving per year by solar assisted hot water system can be 996,772 m.kcal and from hot air systems can be 1,375,908 m.kcal in the year 2000.

In spite of the above mentioned potential, the demand for such devices is directly related to a number of factors that include the following:

(i) Pricing of conventional fossil fuel sources
(ii) Availability of solar systems of reliable quality
(iii) Fiscal incentives
(iv) Awareness of the benefits of such systems in the masses.

ISSUES IN TECHNOLOGY DISSEMINATION

Solar energy systems have been uniformly relevant in the country and also lead to manufacturing and maintenance job creation in different parts of the country. A number of promotional ideas were tried to tap the potential in the country. The march during the decade, since formation of CASE in March 1981, has indicated a number of policy, technology and social issues that need to be tackled for sustained promotion of solar energy devices. These are also sometimes seen as barriers in promotion of solar energy use in the country.

These are as follows:

Policy Issues

The 'Policy Issues' include the following:

(i) Reflection of political will in terms of government's policy that includes legislative measures, financial allocations, and a definite tilt towards greater use of renewable energy sources.

(ii) Long Term Energy Use Plan with intermediate monitoring, evaluation and mid-course corrections to ensure a shift towards renewable energy systems.

(iii) Realisation of environment preserving capabilities of renewable energy systems.

(iv) Realisation of low life-cycle cost of renewable energy supply systems.

(v) Realisation of relevance of decentralised form of energy and, therefore, relevance for integrated rural and urban development in different parts of the country.

(vi) Development of appropriate fiscal package for long term application aimed at promotion of industries, consultancy organisations and users of renewable energy systems.

(vii) Provision of awareness creation through education, information and mass media channels.

(viii) Provision for training courses for manpower development in the field of solar energy manufacture and maintenance.

Social Issues

There are certain important 'Social Issues' which need to be given due consideration. These will include the following:

(i) Realisation of the role of people in policy, implementation and maintenance of equipment and devices. Special concern to be made of women in the domestic sector and for the change in habits and use patterns.

(ii) Realisation of the appropriateness of solar energy devices for certain identified applications to replace conventional sources of energy. This is to be seen as supplementary and not competitive to conventional sources of energy.

(iii) Identifying the possibility of community use for larger dissemination of such systems.

Technology Issues

'Technology Issues' still dominate activities in the solar energy sector as most of the experts and decision-makers belong to the R&D who initiated work 10 years back. Internationally also, the technology issues are given prominence. Following are some of the technology issues:

(i) Development of package of technologies to encompass technical know-how, fiscal incentives, raw material inputs, institutional finance and marketing.

(ii) Creation of appropriate environment whereby research is undertaken only in a few identified specific areas.

(iii) Promotion of industrial R&D as a necessity and a supplementary part to basic R&D in educational institutions. Developing programmes for training on emerging technologies and their maintenance.

(iv) Development of R&D programmes oriented to materials development.

(v) Development of programmes to highlight 'technology-cost' interrelationship to provide most cost-effective total technology package.

10

ROLE OF ENERGY SOCIETIES IN PROMOTING USE OF SOLAR ENERGY IN INDIA

INTRODUCTION

Solar energy application in India are traditionally accepted by the masses in the urban and rural areas since ancient times. Even in the modern times the increasing rural populations and corresponding poverty has made it necessary for the rural populations to opt for cheaper local energy options. Foreign exchange pressure is another element that has necessitated reduction in import of crude oil and use of locally available renewable sources of energy. Such a transformation calls for involvement of large masses in investing their own money for energy processing and energy utilisation devices. Appropriate awareness of such a situation through various formal and non-formal media plays an effective role.

FIRST AWARENESS SEMINAR IN 1982

This was realised way back in 1981 when the Commission for Additional Sources of Energy, CASE, was set up. The first Seminar sponsored by CASE was on 'New Sources of Energy and Mass Media', on March 20, 1982 and I had the privilege to conduct the same under the auspices of the Indian Association for the Advancement of Science.

Experts attending that seminar adopted the following resolution:

"Present energy crisis demands determined efforts from all sections of society to provide alternatives. These additional sources

Address at the UNDP sponsored National Workshop on "Technology Transfer in Solar Energy Applications" organised by the Solar Energy Centre at New Delhi during November 18-22, 1991.

should not be treated as alternative to conventional sources of energy but supplementary to them. This action needs awareness to be created amongst scientists, industrialists, legislators and masses. All possible sources of modern mass communication and traditional communication should be utilised in the common man's language. Various groups of people should continuously interact so that a better understanding amongst them is achieved."

The Seminar also realised "though mass media is not an essential factor in government policy making, it helps to a great extent in making policies socially acceptable. Deeper analysis have shown that the press and the electronic media can have a profound influence on changing the ways of the people only if they are fully supported by the informal channel of communication which are intimately related to basic social processes. Creation of proper awareness calls for coordination and effective use of both mass media and the more personal face-to-face pattern of social communication.

The circumstances in which the mass media is to be utilised for creating an awareness of new and renewable sources of energy is very different and no parallel can be drawn on any of the past practices like propagation of scientific achievements, family planning methods or agricultural practices."

TARGET CATEGORIES

In terms of awareness of energy problems, their course and consequences, two major categories of people have been identified at various platforms. These are people "unaware" of energy crisis, and people "aware" of energy crisis. These are as follows:

(i) The unaware category, by the majority, is the so-called undetected and unenlightened class. This category of people need to be informed about the impact of energy on their existence and are to be made conscious of energy crisis.

(ii) The category of people who are aware, the so-called educated and the enlightened class can be classified as

a) aware but opposed to giving importance to energy consideration;
b) aware but indecisive of reaction;
c) aware but unconcerned; and
d) aware and concerned.

GAP IN INFORMATION GENERATION

The renewable energy programmes were started more as scientific programmes as DNES is an offshoot of DST. Late Dr Atma Ram, a well known scientist and former Director General of CSIR, in his Welcome Address at the above mentioned Seminar said, "How do we organise publicity in a complex problem like energy. It is obvious that the degree of success will depend very much on how it is organised. In my view, good deal of publicity on scientific matters has been either pre-mature or out of proportion and has done more harm than good, though unintentionally to the cause of science and scientists. As I have said on some earlier occasion, one can describe today's situation in the scientific arena as a case of misplaced expectations and over-enthusiastic claims. Even before the projects get off the ground, high expectations are aroused and in six months or a year people begin to expect and ask for the results. For this scientists must share a major part of the blame. It is very essential that there should be some mechanism by which the information given to the public is authentic and is not given in a fashion that creates a difficult situation instead of doing some good."

Incidentally, late Dr Atma Ram was also the President of the Indian Association for the Advancement of Science at the time of making the above statement. His word of caution was based on his life long involvement with the scientific field and the problems faced in application of science and technology—that is the horizontal spread, rather than the vertical growth of excellence in R&D.

Late Dr Atma Ram had foreseen that the renewable energy applications aimed at wider use by the masses have to be disseminated on the basis of the proven technologies and authentic information on the same.

SOURCES OF INFORMATION

In the environment as it exists in the country, the information on renewable sources of energy emerged either from the government machinery (that includes the Central and the State Government), or the government owned educational institutions and national laboratories (as government funding was cautiously provided to such institutions). Evaluation of a process to verify the authenticity of claims cannot be ensured in such an environment. Such claims can at best be discussed, debated and synthesised only on platforms that give professional freedom of questioning. It does not mean that such a freedom will make the project coordinators to give factual down-to-earth information, but will keep them on guard. They can be questioned by an equally competent professional at a certain point of time in the near future. Professional Societies can thus play an important role in this area.

ROLE OF PROFESSIONAL SOCIETIES

Solar Energy Society of India, founded through the concerted efforts of R&D scientists and educationalists, was seen as a platform to discuss such issues regularly at its annual meetings. As it carried the culture of the scientific community, emerging out of the Indian Science Congress Association, the Annual Conventions turned out to be more of a fanfare.

Bio-energy Society of India which started much later has done tremendously well as that is working more as a professional organisation discussing, debating and finalising on the issues important for policy making. The structure of Bio-energy Society of India is democratic in nature and provides equal right and opportunity to every member to question the authenticity of the claims.

There are other professional organisations that have done tremendously well in creating awareness of various issues amongst the concerned groups. To name a few, the Indian Association for the Advancement of Science and the Global Energy Centre have been continuously involved in such programmes.

After a decade of concerted efforts the present emphasis of awareness of energy problems is only related to the people who are aware of the energy crisis. This type of awareness in an authentic fashion can be brought around only by developing an interfacing mechanism that has acceptability amongst the technology generators, policy makers and the users. Such a situation exists only with the professional societies that have an important role to play as they are independent and non-partisan and can objectively study the claims of technology generators and their cost-effectiveness and the energy needs of the people. This information is to be processed and re-packaged for the policy makers to take appropriate action.

The International Solar Energy Society took effective steps and promoted a separate International Association for Solar Energy Education. During the session on solar energy education of North Sun 1988, discussions came up whether there was a need for increased international cooperation in this field. This discussion was followed by an inquiry study which was sent out to the North Sun 1988 delegates and later handed out to the delegates of the 1989 Energy Educators Association Meeting. Of those who are involved in educational activities a great majority were in favour of creation of some sort of International Network.

The International Association for Solar Energy Education was founded with the First Article stating, 'IASEE is a non-profit organisation with the purpose of solar energy education worldwide by means of publications of a Newsletter, organisation of international meetings, and topic activities. Solar Energy is here defined as direct use of the sun's energy as well as other renewable sources of energy.

CONCLUSION

The Solar Energy Society of India and Bio Energy Society of India, in collaboration with other professional societies, can set up a cell that will meet periodically to coordinate education, information and mass media materials in the field of solar energy. To achieve such an object and to disseminate information a periodic Newsletter can be started; or these topics can be included in the existing Journal of the Society. In addition, training workshops and programmes on self-supporting basis may be conducted as is being done by other professional organisations like the Plant Engineers Association, The Institution of Engineers (India) etc. In such a case, SESI will not be always dependent on the government funding as in the case of educational institutions. More and more of such activities can be undertaken only when they are made self-supporting.

The role of the energy societies as an interface is extremely important and relevant.

11

GLOBAL ENERGY PERSPECTIVES AND BIOMASS USE

INTRODUCTION

Sustainable energy supply has an important role in the socio-economic development. Non-availability of supply means absence of energy services which will in turn mean the inability of vast number of people to satisfy their basic needs.

Global energy demand is on the rise to fulfil the socio-economic development objectives. Increasing awareness of the humanitarian consideration will result in heightened awareness of improving the quality of life of the common man, which will again require increased energy consumption levels. On the other hand increasing incomes round the world will also result in people wanting higher levels of better quality, more efficient, cleaner and less obtrusive energy services. Thus, one can really foresee the direction in which the energy consumers are heading. It may be possible to anticipate the forms in which energy will be demanded by consumumers in the future, but it will be difficult to estimate total level of energy demand, or which energy source will meet that demand.

As the Conference is to focus on rural electrification to provide minimum energy needs it would be relevant to outline the issues on rural energy supply. At present the number of people who cannot access commercial energy is a staggering two billion which is roughly 40 per cent of the world's population of 5.8 billion people. The World Energy Council considers the most challenging task to provide energy

---

Paper submitted to the Indo-Italian Workshop on "Electrification through Biomass and other Renewable Energy Sources" organised by the Indian Chapter of International Centre for Theoretical Physics and the Ministry of Non-Conventional Energy Sources at New Delhi during March 20-22,1997

to these two billion people. The 16th Congress of the WEC in Tokyo had specifically recommended as follows:

> "For the 40 per cent of the world deprived of commercial energy services and supporting technology at the present time, effective measures are urgently required to bring commercial energy to them in order that they may break out of the vicious circle of poverty and deprivation".

The World Bank, in its report: "Rural Energy and Development", 1996 has referred to the "nearly two billion people in developing countries (whose) search for energy is a daily grind". They have neither electricity nor gas nor oil to cook their food. The World Bank notes:

> "Women and children from these families often spend hours each day gathering dung and wood—hours they might other wise be able to spend on producing work or education. Their health is damaged by the smoke given off by these fuels. Their environment may be damaged as they chop down trees for wood".

The World Bank also notes that in South Asia traditional biomass accounts for about 55 per cent of total rural energy use; not much more than 25 per cent of rural population, and 50 per cent of urban population are connected to electricity.

The solar initiative of the World Bank embodies the work with its member countries, research and development bodies, industry representatives and NGOs to hasten commercialization of renewable energy technologies and to expand their applications in developing countries.

The Committee on New and Renewable Sources of Energy for Development of the U.N. released a Policy Statement on February 17, 1995 which analysed rural energy needs and prospects in developing countries on the basis of scenarios indicating that the population in the developing world is expected to increase to over 7 billion in 2025, with rural population remaining around 3 billion (as most of the increases in the developing world are to take place in urban centres). Per capita income and productivity in rural areas remain low and poverty is rampant. Income from agriculture is only US $ 300 per capita. Numerous and often grave institutional shortcomings and the lack of basic infrastructure such as energy and water supplies, transport and communications, all contribute to the problems faced by rural areas.

In Africa, Asia and Latin America, 80 per cent of total rural energy consumption is absorbed by households, mainly for the preparation of food; 15 per cent is used for agriculture, trade and industry; and 5 per cent is consumed in transportation. Rural areas in these countries are, for the most part, dependent on draft animal power and traditional fuels such as fuelwood, agricultural and animal wastes. More modern technologies are also playing increasingly important role in some of the developing countries. Coal, crude oil, furnace fuel oil, natural gas and hydropower are used for electricity. It is estimated that during the period 1970-1990 developing countries provided electricity to an additional 1.25 billion in both urban and rural areas. Despite this effort, out of about 4 billion people in the developing world, 2 billion, mostly in rural areas, are still without electricity.

The Committee felt that the challenge now is to develop a strategy to facilitate and expedite the transition towards the use of efficient agro-technologies as well as an appropriate mix of traditional, conventional and renewable sources of energy, in rural communities and economic activities presenting a variety of socio-economic settings. Such a strategy would consist of energy options which would include:

(i) renewable energy sources such as photovoltaic systems, solar thermal energy, biomass energy and hydro power;
(ii) fossil fuels like coal, oil and natural gas.

It was estimated that in 1990 the biomass consumption in developing countries was equivalent to about 900 million tonnes of oil, for a total value of US $ 137 billion which is 1.2 times the total oil consumed in the developing countries in the same year. About 45 per cent of the biomass is wood which is used either directly or in the form of charcoal. Biomass is also used to produce ethanol for transportation fuel. Sustainable biomass energy systems could be the largest single contributor to global energy supply, providing from 17 to 35 per cent of total demand for primary energy in the year 2050. Electricity form biomass sources may also be forming upto 25 per cent of the total electricity generation in 2050.

In this regard the theme of the Indo-Italian Workshop shows the vision of the scientists at ICTP and the Indian Chapter of the ICTP by focussing on biomass in the title of the Workshop as Electrification from Biomass and other Renewable Sources of Energy.

Future projections have been made by different institutions and organisations. However, I will prefer to make a projection on the basis of the findings of the World Energy Council as I had been involved in conducting meetings/studies for the South Asia region. By far these projections have been found useful indicators to future growth patterns.

WEC REPORTS ON GLOBAL PROJECTIONS

The World Energy Council has been making projections for the energy supply and demand at different times. Its Reports titled " Energy for Tomorrow's World" was published in 1993 with the horizon year of 2020. The Report was followed by the WEC-IIASA Report titled "Global Energy Perspectives to 2050 and Beyond" published in 1995. Whereas the first report was based on the deeper studies of different consumption patterns and gave an idea about the possible route for meeting energy needs by the year 2020, the second report mainly referred to the projections as indicators to the year 2050 and beyond.

The paper has drawn widely from the published Reports of the World Energy Council, and discussions at WEC Committee Meetings; The WEC Commission report: "Energy for Tomorrow's World", 1993; the WEC study: " New and Renewable Energy Resources; A Guide to the Future", 1994 and the joint WEC/IIASA study report; "Global Energy Perspective to 2050 and Beyond", 1995.

ENERGY FOR TOMORROW'S WORLD

Energy for Tomorrow's World involved over 500 people from dozens of countries. Regional Working Groups, covering the world, had provided local experience, realism and priorities for the future. The WEC Commission Board comprised of nearly 50 distiguished people from many countries and a variety of disciplines.

As Energy for Tomorrow's World deals with the situation up to 2020, the projections give achievable targets. The forms of energy, technology and speed are more or less foreseeable.

People's Perception and Concern for Energy

The Study focussed on various issues that have direct or indirect bearing on energy consumption in different countries and regions. In their approach the investigators had tried to identify the priorities attached to the key issues on regional basis. Table 11.1 gives an interesting reading.

TABLE 11.1: WEC COMMISSION: REGIONAL KEY ISSUES AND PRIORITIES

	Energy Need & Population	Techno-logy	Fin-ance	Insti-tutioal Defence	Effici-ency & conserva-tion	Climate Change
North America	4	4	4	3	3	3
Latin America	1	1	1	2	1	3
Western Europe	4	3	4	3	1	2
Central & Eastern Europe/CIS	4	2	1	1	1	4
Middle East & North Africa	1	1	1	1	1	4
Sub-Saharan Africa	1	1	1	2	2	4
South Asia/	1	2	1	1	2	4
Pacific of which China	1	1	1	2	2	4

1: Very Important, 2: Important 3: Of Concern, 4: No Concern.

The priorities of different regions in the world differ vastly depending on whether they have plentiful energy at a low cost or a substantial shortage of energy. North America had no concern for energy issues, whereas Africa and Asia-Pacific region attached highest priority.

Demand Projection

The results of the WEC Commission's work in terms of energy demand to 2020 involved projections under three scenarios.

(i) Case 'A' is basically high growth future, in terms of income, energy, and technology;
(ii) Case 'B' has more modest but perhaps realistic growth; and
(iii) Case 'C' assumes aggressive international cooperation focused on environmental protection and international equity.

The WEC Commission had focused on primary energy demand and not focused on final energy. Thus, that could also be taken as indicative to the fact that though the concept of energy efficiency was important, the quality of energy supply was not an important element.

At least 85 per cent of the increase in global energy demand was anticipated to rise in the rapidly industrialising and developing countries. For the regions of Pacific/Asia and South Asia energy demand is expected to double or more by 2020 over 1990 levels in different cases. By the year 2100 global energy demand in the three cases, speculatively extended over that long period, was indicated as rising from its 1990 level of 8.7 Gtoe (gigatonnes oil equivalent) to between 20 and 42 Gtoe.

Effect of the Report

Effect of Projections of the WEC Commission's Report were summarised by Mr. Ian Lindsay, Secretary General, World Energy Council in his presentation at ASIA ENERGY VISION 2020, in November, 1996. He said:

The huge energy growth by 2020 will mean:

(i) More than 90m barrels of oil a day will be consumed, an increase of some 27m barrels per day or the equivalent of the whole of today's OPEC production.

(ii) Coal output will double to about 7 billion tonnes per annum, more than twice the UK's or Canada's known total current coal reserves.

(iii) Gas demand will more than double, to reach some 4 trillion cubic metres per annum, almost as much as the USA's total current gas reserves.

(iv) More electricity generating capacity will be built over the next 25 years than in the previous century.

(v) 90 per cent of this energy growth could take place in the currently defined developing countries particularly in Asia and in Latin America, provided these markets are able to develop satisfactorily with the support and assistance of their authorities.

(vi) Developing countries which today consume 30 per cent of the world's total energy will consume 50 per cent by 2020 and probably 70 per cent by 2100.

(vii) Those same developing countries will emit CO_2 from fossil fuel burning equivalent to the total CO_2 worldwide emissions from the industrialised countries in 1990.

WEC-IIASA STUDY AND GLOBAL ENERGY PERSPECTIVES TO 2050 AND BEYOND

This Report is being dealt at a greater length as this represents the latest trends and a longer time horizon.

The WEC -IIASA report of 1995 projected global energy needs to year 2050 and beyond, using three energy scenarios of the WEC Commission Report of 1993. Each of these was further expanded.

Energy use is the lowest in 'C' of the three cases, but economic and technological growth are higher than in case 'B'. The study Group Leader thought that more variety was needed than was represented by the three basic cases. Case A was therefore, subdivided into three alternative scenarios (A1,A2 and A3); and Case C in two alternatives (C1 and C2). Case B was kept under one scenario.

All the three cases-all six scenarios-start from the base year 1990; and has taken into consideration the historical trends leading up to that year. (Thus it will be observed that both the Reports consider 1990 as the base year). The study looked at the determinance of future energy use, and describes how they vary across the six scenarios. The implications of future energy increases are also reflected in terms of:

(i) investments and financing;
(ii) international trade;
(iii) the constraints and the opportunities facing different energy facilities, - coal, oil, gas , renewable and nuclear; and
(iv) environmental impacts, at the local , regional and global level.

Outlook upto 2020

The study observed that various scenarios look much the same through 2020, and all rely heavily on fossil fuel, but after 2020 the six scenarios start to diverge. Thus, the WEC-IIASA Study further substantiates the findings of WEC Commission Report.

Divergence in Post 2020

Part of the divergence is expected to depend on policy choices and development strategies.

Most of the post-2020 divergence will depend on technological developments and industrial strategies followed by different countries. Following issues will be taken into consideration.

(i) Which energy sources in 2020 will be better matched than others to the more flexible, more convenient, and cleaner forms of energy desired by the consumer?
(ii) Which will have made the investment in research and development that will give them a technological edge?

(iii) Which will have refocused their business from providing just tonnes of coal or kilowatt-hours of electricity, to providing more flexible, convenient and clean energy services to consumers.

The answers to those questions are expected to be determined between now and 2020. Because of long life time of power plants, refineries and other energy investments, there is not a sufficient turnover of such facilities to reveal large differences in our scenarios prior to 2020, but the seeds of the post-2020 world will have been sown by them.

Some Major Conclusions

Mr. Michael Jeferson, Deputy Secretary General, World Energy Council in his presentation at ASIA ENERGY VISION 2020, in November, 1996 presented following as major conclusions of the Study: (detailed observations are given in the Annexure)

— reliance on fossil fuels is likely to remain heavy for many decades.
— geological resources of fossil fuels are plentiful for the next century.
— there is a real risk of growing import dependency and lengthening supply lines for many countries using fossil fuels. For these and other reasons there are uncertainties over the future of fossil fuel prices in the longer term.
— local, regional and global environmental concerns and mitigation measures are likely to lead to widening efforts to incorporate environmental impacts in costs and prices;
— desire to broaden the supply base, build up indigenous capacities, reduce import dependency, and lower atmospheric emissions from fossil fuel use are likely to encourage the expansion of non-fossil fuel supply and use.

Following quantified observations have been drawn : 76 per cent of the world's primary energy is supplied by the fossil fuels. About 12 per cent of the primary energy is supplied by the fuelwood, crop wastes and animal dung. Only 2 per cent is supplied by the new renewables and 4 per cent by large hydros. Thus the new renewables have a very small base to move from and it will take substantial time before they take a prominent place. Refer Table 11.2.

TABLE 11.2: WORLD PRIMARY ENERGY FUEL MIX 1995

Source	per centage Contribution
Coal	23
Oil	33
Natural Gas	20
Nuclear	6
Large Hydro	4
New Renewables	2
Traditional Fuelwood	12

The share of global primary energy consumption accounted for by the present developing countries shift from 34 per cent in 1990 to about 50 per cent in 2020, about 60 per cent in 2050; and to the range of 70 per cent in 2100. (Ref. Table 11.3) Though lack of energy geological resources is not foreseen but there is expected to be growing import dependency on, and Lengthening Supply Lines for, fossil fuels affecting a larger number of countries. This will increase uncertainties about price developments and supply security. For these and other reasons there is expected to be increasing pressure to expand non-fossil fuel supplies.

TABLE 11.3: AREA SHARES OF WORLD PRIMARY ENERGY CONSUMPTION (PER CENT) BY SCENARIO

	1990	2020			2050			2100		
Present		A	B	C	A	B	C	A	B	C
OECD	47	37	38	32	27	28	21	18	16	11
DCs	34	48	49	53	58	60	67	71	74	82
Other	19	15	13	15	15	12	12	11	10	7

The global supply mix by 2050 by various scenarios is diverse. (Ref. Table 11.4)

TABLE 11.4: WORLD ENERGY SUPPLY MIX IN 2050 (Gtoe)

	A1	A2	A3	B	C1	C2
Coal	3.8	7.8	2.2	4.1	1.5	1.5
Oil	7.9	4.8	4.3	4.0	2.7	2.6
Gas	4.7	5.4	7.9	4.5	3.9	3.3
Nuclear	2.9	1.1	2..8	2.7	0.5	1.8
Renewables	4.7	4.9	6.8	3.7	4.8	4.2
Non-Comm.	0.8	0.8	0.5	0.7	0.8	0.8
Total	24.8	24.5	24.5	19.7	14.2	14.2

A similar observation has also been made for the developing countries. (Ref. Table 11.5)

TABLE 11.5: PRESENT DEVELOPING COUNTRIES SUPPLY MIX IN 2050(Gtoe)

	A1	A2	A3	B	C1	C2
Coal	2.1	4.8	2.0	2.8	1.4	1.4
Oil	4.7	2.7	2.6	2.5	1.8	1.8
Gas	2.6	2.4	3.5	2.1	1.9	1.8
Nuclear	0.9	0.5	1.08	1.0	0.1	0.7
Renewables	3.4	3.3	4.7	2.8	3.7	3.1
Non-Comm.	0.7	0.7	0.4	0.6	0.7	0.7
Total	14.4	14.4	14.2	11.8	9.6	9.6

Renewable energy availability (including traditional non-commercial fuels and large hydro) totalled 1.6 Gtoe in 1990. By 2050 under the scenarios that volume is projected to have risen marked in all cases, and by 3 to 4 fold in most scenarios. (Ref. Table 11.6) The pace of development of new renewable forms of energy has recently been slow, mainly for lack of finances. However, the municipal and solid waste disposal will be an area which will receive greater attention for environmental reason. Funding will also come from the city infrastructure development rather than new renewable development.

TABLE 11.6: RENEWABLE ENERGY AVAILABILITY IN 2050 (Gtoe)

	A1	A2	A3	B	C1	C2
Modern Biomass	1.1	2.2	3.2	1.4	1.6	1.5
Solar	1.8	0.4	1.6	0.4	1.3	1.4
Hydro	1.0	1.1	1.1	0.9	1.0	1.0
Wind	0.7	0.6	0.8	0.6	0.6	0.5
Other	0.1	0.6	0.1	0.4	0.3	0.8
Total	4.7	4.9	6.8	3.7	4.8	5.2

ENERGY TRANSITIONS

The two grand transitions in the energy sector have shaped structural changes at all levels. The first of these was the end use innovation of steam engine to be powered by coal. The steam cycle represented the first conversion of fossil energy sources into work; it allowed the provision of energy services to be site independent, since coal could be transported and stored as needed; and it permitted power densities previously only possible in exceptional locations of abundant hydro power. By the turn of the 20th Century, coal had replaced traditional non-fossil energy sources and supplied virtually all the primary energy needs of industrialised countries.

The second grand transition was the greatly increased diversification of both energy end use technologies and energy supply sources. Perhaps the most important single innovation was the introduction of electricity as the first energy carrier, that can be easily converted to light, heat, or work at the point of end use. A second key innovation was the internal cumbustion engine, which revolutionise individual and collective mobility through the use of cars, buses, and aircraft. In particular, oil emerged from being an expensive curiosity at the end of the 19th century to the dominant global position, it has occupied for the last 35 years.

The two grand changes in energy supply were associated with introduction of commercial sources of energy, increasing energy quality and decreasing energy intensity. They also brought around far reaching structural changes in employment and habitation. Industrialisation and urbanisation are attributed mainly to such transitions in energy.

THE NEXT TRANSITION SUITED TO RURAL ENERGY SUPPLY

There is no doubt that electricity will play a dominant role in future development, even of the rural areas in the developing countries. The two billion poor rural people round the globe having no access to commercial sources of energy will no longer have to wait for the commercial sources of energy to come from outside. The locally available renewable sources of energy will have substantially matured technologies and systems that will be locally available.

The next transition in energy is likely to take place by the middle of 21st Century. That will be moving towards use of sustainable renewable sources of energy to fulfil all the energy needs of the increasing population in the world. A major transition since the 1990 has been a focus on the 'quality of final energy' and not on the "primary energy', thereby making it possible to develop technologies for which the starting point is not the main consideration but the end point is the major consideration. Energy supply in the final form is going to be the major consideration. If a desired final energy form can be derived from bio energy source then there is nothing that can stop the development. Future fulfilment of rural electrification is bound to come from the renewable sources as those will be locally available and will be handled with the matured and locally disseminated energy technologies.

There will also be a major transition when the present day government controlled and government funded research, development and demonstration and large scale dissemination will be substituted by private funding and control. As the practice in government planning is in terms of a macro plan and deciding on a national level on the funding pattern, very often the resource crunch becomes a major constraint in the growth of rural energy supply systems. However, a local level plan for energy supply to be funded through private finance stands a better chance for seeing the light of the day. Most of the developing countries have local governance structure which will like this kind of scenario to emerge.

Electricity and liquid form of energy will continue to be the mainstay for the rural areas. Biomass, solar and wind power are three major primary forms of renewable sources of energy that can be favourably and competitively converted into electricity and liquid energy forms. International and regional cooperation in these sectors will grow on the market driven logic and not basically under the bilateral agreements. Transfer of technologies will in future be only through the proven route and not on the basis of the futuristic possibility. The academic and research institutions will first have to prove their technology commercially in cooperation with industries of the country of origin on the scale at which they wish to transfer the technology before really taking up any such transfer of technology.

The future options will emerge with wider participation of the user agencies and with larger financial contributions from communities and the users with governments playing the role of a facilitator only. The international funding agencies will become an important actor on the scene but will like to fund only such projects that can be projected to be commercially viable. In every case it will not be possible for the government to dictate their social or economic priorities to the international funding agencies, rather the international funding agencies will be dictating the governments as to where they should shift their social and economic priorities.

CONCLUSION

The future energy course will also be dictated by the environmental concerns and reduction in pollution levels. It may be envisaged as a difficult proposition but it is at present a desirable requirement which will become an essential requirement in time to come. The likely pollution liability will become a major constraint in introduction of a energy conversion or end use device. This again indicates that the renewables will have a greater role to play as a source of locally available primary energy. There are expert opinions indicating that nuclear energy will also play an important role but only after the development and introduction of smaller size nuclear power plants in the range of 125 to 150 MWe fully matured systems are put in operation. Common man's apprehensions on safety over the total fuel cycle will be the major deciding factor. The final energy mix may come from a combination of nuclear and biomass fuels, which is sometimes termed as " Bio-Nuc Fuels".

Technology push and policy pull will be the basic elements for success. Fundamental shifts in policies in the energy sector, requisite flows of investment funds mobilisation, market and regulatory reforms implementation, cooperation and national, regional and international levels are prerequisites for rapid economic and energy developments. The current practice of R&D experts being the link between the market driven industry and the government has to be strengthened.

International Centre for Theoretical Physics has an important role in making this transition a success. With their close interaction with the governments and the industry and the high credibility the scientists have a chance to influence the technology development with the industry and policy making of the governments to ensure a sustainable energy supply system based on the renewables, mainly the biomass.

ANNEXURE

SUMMARY OF RECOMMENDATIONS

Major observations of the WEC/IAASA Study can be summarised as below:

It is relevant to mention that 11 different regional groups are already working on revising the perspectives as experts from different regions have given their impressions that are in some cases at variance from the Study. An in depth analysis is likely to refine the projections.

(i) *World Energy needs will increase:* The world population is expected to double by the middle of the 21st century, accompanied by economic development. The likely result is a three to five fold increase in world economic output by 2050 and ten to fifteen fold by 2100. Global demand for final energy services will grow by as much as an order of magnitude by 2050. Primary energy demand will grow less due to improvement in energy intensities. The expected increase in primary energy requirement is 1.5-3 fold increase by 2050 and 2 to 5 fold increase by 2100. While making these projections the global commercial energy consumption per capita is assumed to increase annually in the period 1990-2020 by between 1.4 per cent and 1.7 per cent. the rate of increase tends to rise to between 1.3 per cent and 2.0 per cent in the period 2020-2050.

(ii) *Energy Intensity will Improve Significantly:* Rapid industrial growth will be accompanied by replacing inefficient technology by more efficient technology thereby resulting in lesser amount of primary energy needed per unit of GDP. The average global reductions in energy intensity range from 0.8 per cent per year to 1.4 per cent per year. Some regions are expected to reach even 2.7 per cent per year.

(iii) *Resource Availability will not be a Major Global Constraint:* The indications are that the economic development over next century will not be constrained by geological resource scarcities. Regional shortages and price increases can occur , due to the unequal distribution of fossil resources, but globally, they are not a constraint. The environmental constraints, financing and technological needs appear a more likely source of future limits and international politics may improve from time to time.

(iv) *Capital Requirement will Present Major Challenges for All*

Energy Strategies: For all scenarios the capital requirements of energy sector are extremely large and the investment are likely to expand at a slower pace than overall economic growth. However, the energy sector will have to raise an increasing fraction of its capital from the private sector, where it will face stiffer competition and return on investment criteria rather than it has in the past. Most of the investment needs will be in the South.

(v) *Energy End-Use Pattern will Converge:* The historical drive toward ever more flexible, convenient and clean fuels reaching the consumer continues in all six scenarios. High quality energy forms become transitional fuels of choice. Final energy structures are converging in all six scenarios, while primary energy structures are diverging. No matter what the eventual dominant fuel scenario, there is shift away from non-commercial and mostly unsustainable uses of biomass; and direct uses of coal virtually disappear. While fossil fuels continue to provide most of the world's energy up to the first quarter of the next century; sustainable uses of renewables come to hold a prominent place in all scenarios, reaching the consumer as electricity, liquids or gases, rather than as solids in the case of biomass.

(vi) *Quality of Energy Services and Forms will Increasingly shape Future Energy Systems:* The energy system is service driven from the consumer to producer, while energy flows are resource and conversion process driven, from the producer to consumer. The Scenarios demonstrate the need to consider end-use and energy supply simultaneously, both from an analytical as well as a policy perspective.

(vii) *Local Environment Impacts will take Precedence Over Global Change:* The natural capacity of the environment to absorb higher levels of pollutions, particularly in densely populated metropolitan areas, will become the limited factor for the unconstrained use of fossil fuels. In the developing world an urgent environmental problem is related to indoor pollution. Both cleaner fuels and active abatement measures will be required. Original air pollution could prove also problematic, especially in the rapidly growing, densely populated coal intensive economies of Asia.

(viii) *Decarbonisation will Improve the Environment at Local, Regional and Global Levels:* The continuing shift to higher quality fuels means continued decarbonisation of the energy system. Decarbonisation on the other hands means, lower adverse environment impacts per unit of energy consumed, independent of any active policies, specifically designed to protect the environment. Restrictions on nuclear power, for example, mean a possible greater dependent on fossil fuels and vice versa. In all cases however, more rapid technology improvement means quicker progress towards the clean fuels desired by consumers and cleaner fuels means a cleaner environment.

12

GLOBAL BIOMASS PERSPECTIVE

INTRODUCTION

Biomass energy has traditionally been associated with the energy supply only in the developing world. FAO has observed that around 5 per cent of the world's energy supply is derived from wood (this 5 per cent also is equivalent to about 475 million tonnes of oil). While globally the fuelwood contributes only 5 per cent of primary energy supply, it is an important source of energy in developing countries. It contributes 15 per cent of their total consumption, and in developing Africa (the most dependent region) it averages 58 per cent of consumption. The growth of consumption of wood fuels worldwide has averaged about 2 per cent per year over the past 20 to 30 years, with virtually zero growth in developed countries, but somewhat over 2 per cent in developing countries. The trend in the developing region has been closely related to population growth, but the growth in future years may be expected to be below the rate of growth of the population. As income level rises there is an increasing preference for more convenient alternative fuels. The World Energy Council, in its projection to year 2020 has, however, reflected that till 2020 the energy consumption pattern will be more or less directly proportional to the population increase. The present day technologies will continue to play their important role and it will only be after 2020 that the effect of emerging technologies will be felt and energy consumption growth may delink from population growth rates. By 2020, many current energy end-use devices will have been replaced by those just

---

Key Note Address at the annual Convention of the Bio Energy Society of India organised at the Indian Institute of Technology, Delhi on 15-16, 1996 at New Delhi.

being introduced for commercial use today: new vehicles, new industrial processes and new systems. Many power plants will have been replaced and others will be nearing end of their useful life. The power horizon to 2050 and beyond means that all energy technologies and devices are likely to have been replaced at least twice.

BIOMASS ENERGY CAN ALSO SUPPORT HIGH GROWTH PATTERN

Energy for Tomorrow's World, the World Energy Council's Report, published in 1993, forecasting energy situation upto 2020, indicates that "energy use (until 2020) will rise and the world will have to rely largely on fossil fuels."

The World Energy Council in its high growth technology intensive scenario (one of the three scenarios) has considered increased role of renewables—especially the biomass. This appears attainable and most likely emerging scenario. The concept of this Paper is already knit around this hypothesis. 'new renewables' and 'nuclear' combine to form a 'Bio-Nuc' technology cluster that terms the transmission to a post-fossil age along similar market place dynamics by which fossil fuels phase out traditional energy forms in the course of the 19th century. Natural gas provides the transitional fossil fuel of choice. This strategy leads to a significant degree of decarbonisation of the global energy system. Global carbon emissions are projected to amount to 9 GtC in 2050 compared to 6 GtC today, while primary energy use is projected to increase of 25 GtC, which is 3 times higher than at present. It is an illustration of a case in which our rich and clean energy future can overcome some of the challenges of global warming without recourse to stringent environmental policy measures.

Development pressure of the developing economies will be the key factor in deciding energy consumption pattern. Some of them being termed as Asian Tigers have already indicated a rapid growth, but this rate of growth is not likely to be sustained over a longer time horizon. Wherever the rates of growth will be higher that will be accompanied by higher technology interventions and energy efficiencies and cleaner environmental technologies. The finance for them will be a constraint because of their own in-built capabilities. Most of these fastest growing economies have their own sound basis of fossil sources of energy or have developed a strong export market so that they generate sufficient trade surplus to meet energy imports. The story does not end at this point. The concern for the future would largely be for the growth and development of the people from the developing world, i.e., basically from the Sub-Saharan Africa and the South Asia regions. Their interest normally does not figure in global reflections. They will have the need to enter the high growth pattern but will not have sufficient access to fossil sources of energy to meet this scenario of growth and will thus have to depend largely on the technology intensive growth pattern. The economic growth needs and social development needs will have to be segregated. The fossil energy meeting economic growth needs, and renewables meeting the social development needs.

GEO-POLITICS OF ENERGY SUPPORTS LARGER ROLE FOR BIOMASS

Also at the 16th Congress of the WEC, in his Global Energy Address, Mr Andre Giraud, former Minister of Energy and Industry of France and President and Director General of COGIDEV, France, elaborated on the situation likely to control the growth of energy consumption over the next 25 years. He indicated that in 25 years the geo politics of energy can easily undergo a major transformation in supply of energy in the structure of supply and consumption and the configuration of positions of strengths and weaknesses. The geo-political developments from now to year 2020 will be controlled by three important factors which will interact with each other: technical progress, demographic compulsions of the world populations and their standards of living; and heightened perception of the need to respect the conditions of life on the planet.

Mr Giraud reflects that it is not inconceivable that the period in question will see the patterns of major new energy sources, in the forefront of which is biomass from solar energy conversion, that is biological fuel complementing fossil fuels. He said: "I am aware that there is much skepticism here, and that apart from exceptional cases, the cost price of biomass fuels today is 3 times higher than that of traditional fuels. This is also what was said about nuclear energy less than 5 years before it became competitive. If there is a science which is making rapid progress at this moment, it is Biology: Plant Biology to change the solar cultures which plants are, by technologies to ensure transformation and refining. In my opinion, it is only a matter of time when biomass enters the competitive arena.

The geo-political reflections could be considerable because the areas favourable to biomass production are not the same as those which produce traditional sources of energy. The methods of exploitation are not at all the same either, being most suited to decentralisation and the use of human resources. Consequently, we can anticipate a new sector—agro energy.

With the help of technical progress, we can assume that the energy requirements of humanity will be met overall, with more respect for the environment, and at prices which will not displace much from the situation we know today, leaving aside any political crises which may arise."

SUPPORT FOR HI TECH DEVELOPMENT IN EUROPE FOR BIO FUELS

A changeover from centralised to decentralised systems will no doubt cause many a problem. But, in countries like India, that practice decentralised functioning, the biomass energy route has the greatest promise.

Two major issues that Mr Giraud has pointed out are basically relating to 'the environmental necessity' and 'the depleting fossil sources of energy' But he has indicated a very positive shift towards the biomass energy.

Very few people realise that Europe has already entered the area of bio-energy exploration in a very big way. Starting from a modest Expert Group Meeting at the FAO Headquarters in September, 1992, Bio-Fuels are now centre-stage. The European Bio-Energy Society which was a low key activity has been energised. From 1993 onwards a number of discussions on incorporation of emergent conversion and combustion technologies for biomass sources have taken place. Large funds have been allocated for intensive and extensive R&D activities under the European Community, with first initiative coming from France. Realising the dimension of R&D support required, the European Union has supported multi-country funded programmes. Dr Wolfgang Palz, who was once a proponent of solar photo-voltaics, is this year the Chairman of the Organising Committee of the European Biomass Congress starting in Copenhagen on June 24,1995. Such a massive transformation of hearts in people practicing in market economy cannot be just for a social cause. It is all guided only by the potential markets opening up in the field of biomass energy.

Major UN agencies like UNIDO, FAO, UNESCO and UNDP are also orienting their programmes and schemes to support use of biomass as a source of energy. These are a few initial indicators to the fact that biomass is no more a source of energy only for poor in the developing countries. The technological interventions of modern energy conversion process is a major breakthrough due to which bio-fuels are gaining importance in the energy scene. As technologies for conversion of biomass develop and mature the financial projections will also become favourable for utilisation. Initiative of the European Union will certainly be backed up by funds.

FACTORS ENCOURAGING SHIFT TO BIOMASS ENERGY

The most significant factor in the energy scenario over the next 25 years will be the 'shifting populations within a country' and not only of the 'increasing populations'. By this, I mean a shift from rural to urban areas and along with that carrying the habits of the rural masses to the urban centres. The mental blockage of the urbanites towards biomass fuels being a secondary source of energy will be overcome by the rural influx and neo-urbanites. Recent Habitat-II Conference, at Istanbul, during June 3-14, took into consideration the shifting patterns of habitation. It was observed that the problem is getting serious in the developing countries as people move away from rural areas to the urban areas. Cities of developing countries account for almost 50 per cent of the populations and 50 per cent to 80 per cent of Gross National Product. This trend is likely to continue unabated. Overall 80 per cent of the world's urban residents will live in developing countries by 2025.

Some of the emerging indicators need to be referred to much more seriously. Most poor areas of Sub-Saharan Africa and South Asia are expected to see the largest increases in population during the period 1990 to 2020. Sub-Saharan Africa is likely to increase its population from 502 million to 1195 million and the South Asia is expected to increase its population from 1156 million to 1938 million. These increases of 693 million and 792 million in themselves are substantially detrimental to the growth pattern.

Waste disposal, environmental pollution and energy availability are three major components that have to be seriously considered by the urban planners. These three issues are inter-related. Therefore, future efforts of energy in urban areas need be through the biomass route as far as possible.

There is likely to be a significant expansion of renewables. Traditional uses of renewables and traditional technologies are gradually phased out. Old renewables are gradually replaced by commercially high quality energy carriers, including those from new renewable sources. End use conversion of traditional biomass, however, improves significantly and also has the added benefit of reducing deforestation rate and improving the air quality. Attention of energy laboratories on biomass energy R&D has developed hi-tech end processes that are likely to change the shape of things in the future. The production of biomass route is more favoured and is already promoted in Europe, with efforts to produce liquid fuels.

PRODUCTION RESOURCE HOLDS THE KEY

The land requirement for biomass production for an illustrative high growth rate scenario can be studied further. Depending on productivity increases the land requirement would range between 390 and 690 million hectares by 2020. These additional land areas for both agricultural and biomass growing are in principle available, in particular in developing countries, where most of such land requirements are concentrated. No doubt, they stretch future land requirements to their ultimate limits. Both agricultural and biomass energy are moving into a high tech, high productivity future where eco systems are tightly managed and controlled by man.

I do not wish to say that the land mass be taken out of agriculture and put to additional biomass energy crops. Experiments have shown that multiple cropping systems, where agricultural crops and the energy crops are grown side by side, are already proving financially viable; and the farmers have started adapting them. The contribution of biomass energy grown in a planned manner has a significant role to play at the micro level energy scenario. Examples have been observed in India, where Wasteland Development Schemes in the State of Haryana, after three years of operation, have resulted into stabilising the fuel wood trading price and in some cases even bringing down the price in the local market. This is observed, particularly in a fuel wood deficient area, where the prices in the past have always been going up the spiral.

CONCLUSION

Whether it be for population increase or even for the lack of fossil resources, biomass energy is going to play a significant part in the years to come, as the rural populations move on to the urban centres. However, its appropriate exploitation for high quality energy will require appropriate technology development and its dissemination in a uniform fashion.

The growing global energy needs will be accompanied by the energy intensity that will improve significantly due to improvement in process technologies, conversion processes and end-use devices.

Resource availability for 'final energy' will not be a constraint, although from the primary energy point of view it will be a major constraint.

The present trends indicate that the energy transition has the strongest possibility of moving along the third scenario under the high economic growth reflected in the Report of the World Energy Council presented at the 16th Congress at Tokyo. To achieve that will require much greater involvement from the society and much greater intervention of the emerging commercial technologies so that maximum benefit is derived by converting biomass sources in the most efficient fashion. Work on the energy efficiency sector will have the largest contribution in promoting biomass sources of energy in the rural and urban areas. Adoption of all modern technologies of conversion will really hold the key for making the biomass the Number One fuel over next 25 years.

13

ENERGY FOR SUSTAINABLE DEVELOPMENT OF THE MILLIONS

An Overview of Energy Situation in the Commonwealth

INTRODUCTION

50th Anniversary celebration is a landmark in the growth and development of the Commonwealth Engineers' Council. Befitting to the occasion, Her Majesty the Queen of Britain, met each one of the participants of the International Conference on "Engineering for Survival" in London in March 1996. Her discussions were centred around the concern for the common man, especially the poor. She obviously wanted engineering skills to be used for the uplift of society. I was deeply impressed and touched by her humility.

The International Conference at that time discussed various aspects of sustainable development and the role of engineering skills in attaining the same. India being one of the five founder signatories to the Commonwealth Engineers' Council was given the responsibility of discussing "Energy for Sustainable Development of Rural Areas." The Indian team led by the President of The Institution of Engineers (India) worked on the title to truly reflect the spirit in which the Conference was to be planned. "Energy for the Millions" was considered the most appropriate title to indicate the concern for the poorest of the poor. The aim obviously was chalked out to bring out effectively the role of energy in meeting all the needs for social and economic development. This included energy for minimum needs like cooking; income generation activities like agriculture and small scale

---

Paper submitted to the Commonwealth Engineers Council, CEC, sponsored International Conference in "Energy for the Millions" at Bangalore during December 19-20, 1996.

industry; and area development schemes. Sustainable development was considered necessary—and that could be attained through self-supporting energy supply systems. Therefore, the initial thrust has to be on "energy for income generation" so that vibrant decentralized development of rural areas results in higher purchasing capacities through higher generation of wealth. However, such a development was considered possible only through sufficient funding from the government as well as private sources. Pricing of energy being a major issue for long term sustainable energy supply.

THE COMMONWEALTH GROUP

The Commonwealth represents almost 1/3rd of the world's nations having diverse socio economic and cultural systems and values. The idea of a Commonwealth of Nations comprising Great Britain, the Dominions and other territories in the British Empire was first accepted at the Imperial Conference of 1926. In 1931, the statute of Westminster recognized the status of the Dominions and defined the relations of the British Crown to the dominions. The other territories were entitled to become members of the Commonwealth on attaining full self government. In 1947, the office of the Secretary of the State for Dominions was abolished and the Secretary of Commonwealth Relations assumed charge.

Some of the members like Canada, Australia and New Zealand recognized the Queen as the titular Head of the State and have Governor General appointed by the Queen on the recommendation of the State Cabinets. Other countries like India and Sri Lanka, who have elected Presidents of their own as Heads of State, recognized the queen as the Head of the Commonwealth only. The Commonwealth Heads of Government Meet (CHOGM) has become an important international event and discusses issues of common interest to the Commonwealth Group of countries. They have also shown keen interest in the energy sector by establishing the Commonwealth Regional Consultative Group on Energy in February, 1978.

ENERGY RESOURCES AND PRODUCTION PATTERNS

An overall review of energy availability, both in terms of conventional sources of energy and non-conventional sources of energy indicate rich endowments in the Commonwealth group. The small island countries, however, have experienced serious and difficult situations and call for specific remedies. A consideration on resourcewise energy supply in the Commonwealth gives the following scenario:

Coal

Coal has been an important source of primary energy in most of the populated Commonwealth countries. Australia has rich coal resources and the high quality of coals are found in Queensland and New South Wales. There are economically viable resources of bituminous coal. Bangladesh, Botswana, Canada, India and U.K. also have considerable coal resources in the Commonwealth Group. Coal deposits contribute significantly to the export earnings of Canada and Australia. Mines in most of the countries are privately owned. India has the largest coal production of all the Commonwealth countries. Coal is expected to be a stable source of energy for a number of years to come. Refer to Tables 13.1, 13.2 and 13.3 for coal reserves, production and consumption data.

Crude Oil and Natural Gas Liquid Resources

Crude oil and natural gas liquid resources have an important role to play in the Commonwealth countries, but their availability is mostly constrained. Australia has considerable crude oil and condensate production programme. Bangladesh also has gas and condensate producing wells, although small quantity of oil is also produced. Brunei is South East Asia's third largest oil producer, with 7 off-shore and 2 on-shore fields, with nearly 60 per cent of production derived from two of the off-shore fields. Brunei is the youngest of the Commonwealth group to become an oil producer in 1991 and an oil exporter in 1992. Canada has strong oil and gas sector. It is expanding its exploration activities into the southern parts in the off-shore area, closer to the North Pole. India also has substantial proved recoverable reserves of oil, sufficient to last for 25 years at current levels of extraction. India is evolving a policy to restrict oil imports at a reasonable level. Meeting the oil requirement is still mainly dependent on oil imports. Malaysia is self-sufficient in oil and is also a major exporter; most of its production being tapped from off-shore oil fields. Crude production has seen a decline after the 1992 levels as its major state owned oil company Petronas continued its emphasis on gas development. New Zealand has off-shore fields continuing to dominate development plans. Pakistan's own production supplies only 1/3rd of the country's growing internal needs and is looking out for attracting larger foreign investments. For small island country of Trinidad and

TABLE 13.1 Proved recoverable reserves of coal-Commonwealth

	Bituminous including anthracite	Sub-bituminous	Lignite	Total
AFRICA				
Botswana	3,500	-	-	-
Nigeria	21	169	-	190
Swaziland	116	999	-	1 115
Tanzania	200	-	200	-
Zambia	-	55	-	55
Zimbabwe	734	-	-	734
AMERICA, NORTH				
Canada (a)	4,509	1,287	2,827	8,623
ASIA				
India (a)	68,047	-	1,900	69,947
Malaysia	4	-	-	4
Pakistan (a)	-	734	-	734
EUROPE				
United Kingdom (a)	2,000	-	500	2,500
OCEANIA				
Australia	45,340	3,700	41,900	90,940
New Zealand	27	81	9	117
Total	124,498	7,025	47,136	175,159
World Total	5,19,358	1,97,096	3,15,156	1,031,610
% of World Total	24	4	15	17

Note : All data in Table refer to the Data Base of the World Energy Council and refers to the year 1993.

Tobago oil export represents approximately 57 per cent of earnings. United Kingdom is a major oil producing country in the Commonwealth group. Exploitation of oil and gas from the continental shelf under the North Sea provides self-sufficiency in energy and has also made the United Kingdom the most energy rich of the European Union Nations. Refer to Tables 13.4, 13.5 and 13.6 for crude oil reserves, productioin and consumption patterns.

Natural Gas

Australia has a well developed North West Shelf Project. Bangladesh has well developed east belt region for gas production. Brunei finds itself well placed with reference to gas production and utilization and

TABLE 13.2: Production of coal (MTonnes) - Commonwealth

	Bituminous including anthracite	Sub-bituminous	Lignite	Total
AFRICA				
Nigeria	0.1	-	-	0.1
Swaziland	0.05	-	-	0.05
Zambia	-	0.4	-	0.4
Zimbabwe	5.27	-	-	5.27
AMERICA, NORTH				
Canada	35.31	23.66	10.04	69.01
ASIA				
India (a)	246	-	17	263
Malaysia	0.26	-	0.03	0.29
Pakistan (a)	-	3.3	-	3.3
EUROPE				
United Kingdom	67.46	-	-	67.46
OCEANIA				
Australia	158.47	18.06	47.65	224.18
New Zealand	1.22	1.7	0.18	3.1
Total	514.14	47.12	74.9	635.8
World Total	3,169	374	931	4,474
% of World Total	16	13	8	14

Note: All data in Table refer to the Data Base of the World Energy Council and refers to the year 1993.

most of its production is shipped to Japan. Canada's gas reserves are now declining and 60 per cent of its gas reserves are concentrated in Alberta. Approximately half of all Canadian production is exported to United States. India has also launched a major programme for utilization of natural gas for the internal consumption. Malaysia has recently discovered higher availability of gas reserves. New Zealand has been producing and utilizing gas since 1979. Pakistan is building plans to use its natural gas to meet 1/3rd of the country's growing internal energy needs. Production of gas in Trinidad and Tobago has been steadily rising with the foreign interest in the same. The United Kingdom has output and reserves of natural gas which are playing a growing role in the country's energy plans. Output and reserves of natural gas are both increasing rapidly.

Refer to Tables 13.7, 13.8 and 13.9 for natural gas reserves, production and consumption patterns.

Nuclear Power

There has been no fundamental changes in the worldwide development of nuclear power in the recent years. Most countries with nuclear programmes followed established plans to expand nuclear power's contribution to their electricity production. Commonwealth also has substantial interest in nuclear power. Canada's nuclear programme is based on the heavy water natural uranium grid system which enables Canada to make use of Canadian reserves and technology. Nuclear power has become important in Canada because in some regions (particularly Ontario) major hydro resources have already been largely developed leaving little scope for expansion; and fossil fuels would have had to be imported. In 1994, about 19 per cent of total power generation in Canada came from 22 Candu reactors. India has nine nuclear units in operation with a total capacity of 1.5 GW(e) contributing less than 2 per cent of its power generation. India now proposes to raise its nuclear power generation capacity to 20,000 MW(e) by the year 2020. Pakistan has a 125 MW(e) pressurized water reactor. The national policy is to expand the share of the nuclear power in electricity generation and, therefore, another 300 MW(e) pressurized water reactor has been contracted with China. United Kingdom has 35 units in operation, with a total capacity of 11.9 GW(e), providing 26.7 per cent of electricity generation. All countries in the Commonwealth have now taken to support only commercially viable nuclear power generation systems in their countries and are working on techniques of safe disposal of nuclear waste. Refer to Table 13.10 for nuclear capacity and actual generation.

Hydro Power

Hydro power is a major source of electricity generation in the Commonwealth countries. 62 per cent of Canada's electricity generation comes from hydro power. Canada has a significant capacity of small hydro plants. India has 27.8 per cent of electricity generation from the hydro resources in the year 1991-92. Serious and continuous efforts have seen emergence of small and mini hydro systems upto the capacity of 15 MW. With the government support in terms of subsidy for capital costs for civil works and equipment, including generator and transformers, and soft loan from the Indian Renewable Energy Development Agency, to private sector entrepreneurs, the possibility of small hydro coming out prominently cannot be ruled out. New

TABLE 13.3: Consumption of coal (Mtonnes)-Commonwealth

	Bituminous including anthracite	Sub-bituminous	Lignite	Total
AFRICA				
Mauritius	0.06	-	-	0.06
Nigeria	0.06	-	-	0.06
Swaziland	0.07	-	-	0.07
Zambia	-	0.4	-	0.4
Zimbabwe	5.39	-	-	5.39
AMERICA, NORTH				
Canada	15.85	23.65	10.01	49.51
Dominican Rep.	0.12	-	-	0.12
Jamaica	0.06	-	-	0.06
ASIA				
Bangladesh	0.38	-	-	0.38
Cyprus (a+)	0.11	-	-	0.11
India(a)	256	-	17	273
Malaysia	2.06	-	-	2.06
Pakistan (a+)	4.24	-	-	4.24
EUROPE				
Malta (+)	0.3	-	-	0.3
United Kingdom	86.8	-	-	86.8
OCEANIA				
Australia	33.87	18.06	47.65	99.58
New Zealand	0.43	1.7	0.18	2.31
Total	405.8	43.81	74.84	524.45
World Total	3,255	377	933	4,564
% of World Total	13	12	8	12

Note: All data in Table refer to the Data Base of the World Energy Council and refers to the year 1993.

Zealand's 68 per cent of electricity generation in 1993 was from hydro power. 45 per cent of Pakistan's electricity is generated from hydro power. Papua New Guinea generates 5 per cent of electricity from the hydro power, but is likely to increase the role of hydro power in the coming years in electricity generation. 94.5 per cent of electric power generation in Sri Lanka comes from hydro power. Refer to Tables 13.11, 13.12 and 13.13 for hydropower capacities and installed capacities for schemes more than 10 MW and less than 10 MW.

TABLE 13.4: Proved recoverable reserves of crude oil and natural gas liquids - Commonwealth

	Crude Oil (m tonnes)	Natural Gas liquids m tonnes	Total (m tonnes)	Total m barrels (app)	R/P Ratio Year
AFRICA					
Ghana	N	-	N	0.5	>100
Nigeria	1,409	284	1,693	12,449	18
AMERICA, NORTH					
Barbodos	0.4	-	0.4	2.9	6.3
Canada (a)	582	176	758	5,574	10
Trinidad & Tobago	69	4.5	73	537	11
ASIA					
Bangladesh	0.5	-	0.5	3.7	18
Brunei	184	-	184	1,353	23
India (a)	757	19	776	5,706	29
Malaysia	585	-	585	4,301	19
Pakistan (a)	27	-	27	199	9.3
EUROPE					
United Kingdom(a)	605	-	605	4,449	6.4
OCEANIA					
Australia	117	82	199	1,463	10
New Zealand (a)	17	-	17	125	9.3
Papua New Guinea	31	-	31	228	5.3
Total	4,383.9	565.5	4,948.9	36,391.1	
World Total	138,287	2,390	140,676	1,034	384,47
% of World Total	3	24	4	4	

Note: All data in Table refer to the Data Base of the World Energy Council and refers to the year 1993.

Fuelwood

While, globally, fuelwood contributes only 5 per cent of total energy supply, it is an important source of energy in developing countries. It contributes 15 per cent of their total consumption, and in developing Africa averages 58 per cent of consumption. The trend in developing regions has been closely related to population growth, but the growth in future years may be expected to be below the rate of growth of population. As income levels rise, there is an increasing preference for more convenient alternative fuels, at the same time, population increase results in deforestation in other areas, a diminution in supply

TABLE 13.5: Production of crude oil and natural gas liquids-Commonwealth

	Crude Oil (m tonnes)	Natural Gas liquids m tonnes	Total (m tonnes)	Total m barrels (app)
AFRICA				
Nigeria	94.83	8.26	103.09	758
AMERICA, NORTH				
Barbodos	0.06	-	0.06	0.4
Canada (a)	73.3	17.2	90.5	665
Trinidad & Tobago	6.4	-	6.4	47.1
ASIA				
Bangladesh	0.03	-	0.03	0.2
Brunei	7.87	-	7.87	57.9
India (a)	27.03	-	27.03	199
Malaysia	30.58	-	30.58	225
Pakistan (a)	2.94	-	2.94	21.6
EUROPE				
United Kingdom	94	6.1	100.1	736
OCEANIA				
Australia	20	3	23	169
New Zealand(a)	1.87	0.02	1.89	13.9
Papua New Guinea	5.9	-	5.9	43.4
Total	364.81	34.58	399.39	2,936.5
World Total	2,997	182	3,179	23,371
% of World Total	12	19	13	13

Note: All data in Table refer to the Data Base of the World Energy Council and refers to the year 1993.

results in higher supply costs, further straining consumption growth.

Fuelwood remains a major source of domestic cooking energy supply for the rural areas of India, Sri Lanka, Tanzania, Nigeria, Ghana, Bangladesh and Pakistan. Though there had not been programmes aimed at fuelwood generation, the dead wood from the forests are the main source in these countries for the rural domestic energy supply. However, over the number of years research efforts have made it possible to develop plant species that can give a high yield for energy plantation crops. India has a national policy for use of wood for energy purposes and has aimed at harvesting fuelwood resources. That means energy plantations are being promoted in addition to timber plantations. Such plantations are being promoted

TABLE 13.6 Consumption of crude oil and natural gas liquids-Commonwealth

	Crude oil Mtonnes	Natural gas liquids Mtonnes	Total Mtonnes	Total Mbarrels (approximate)
AFRICA				
Ghana	0.99	-	0.99	7.3
Kenya	2.27	-	2.27	16.7
Nigeria	11.68	0	11.68	85.9
Sierra Leone	0.29	-	0.29	2.1
Tanzania (a)	0.58	-	0.58	4.3
Zambia	0.55	-	0.55	4
AMERICA, NORTH				
Barbodos	0.27	-	0.27	2
Canada (a)	65	16.7	81.7	600.7
Dominican Rep.	1.95	-	1.95	14.3
Jamaica	2.06	-	2.06	15.1
Trinidad & Tobago	5.41	-	5.41	39.8
ASIA				
Bangladesh	1.1	-	1.1	8.1
Brunei	0.24	-	0.24	1.8
Cyprus	0.78	-	0.78	5.7
India (a)	54.3	-	54.3	399.3
Malaysia	10.79	-	10.79	79.3
Pakistan (a)	12	-	12	88.2
Singapore	51.82	-	51.82	381
Sri Lanka	1.87	-	1.87	13.8
EUROPE				
United Kingdom (b)	85.2	1.5	86.7	637.5
OCEANIA				
Australia (b)	33.93	*	33.93	249.5
New Zealand (a,b)	4.23	N	4.25	31.3
Total	347.31	18.2	365.53	2,687.7
World Total	3,101	47	3,148	23,150
% of World Total	11	38	12	12

N less than 0.1

Note : All data in Table refer to the Data Base of the World Energy Council and refers to the year 1993.

TABLE 13.7: Proved recoverable reserves of natural gas-Commonwealth

	10^9cubic metres	10^9cubic feet	R/P ratio years
AFRICA			
Ghana	23	800	>100
Namibia	147	5,191	>100
Nigeria	3,451	121,872	>100
Tanzania	116	4,097	>100
AMERICA, NORTH			
Barbodos	0.1	4.9	7
Canada	2,232	7,823	16
Trinidad & Tobago	232	8,193	28
ASIA			
Bangladesh	370	13,067	60
Brunei	396	13,985	40
India (a)	686	2,226	34
Malaysia	2 150	75,927	86
Pakistan (a)	646	22,821	39
EUROPE			
United Kingdom	610	21,542	8.6
OCEANIA			
Australia (a)	555	19,600	22
New Zealand	85	3,002	13
Papua New Guinea	425	15,009	>100
Total	12,215.1	335,159.9	
World Total	141,335	4,991,240	81
% of World Total	9	7	

Note: All data in Table refer to the Data Base of the World Energy Council and refer to the year 1993.

preferably on wastelands to achieve the dual objective of restoring wasteland and providing fuelwood for cooking purposes. Continuous efforts at improving the efficiency of traditional wood supplies has resulted in reducing fuelwood requirement in many of the countries. Refer to Table 13.14 for forest area and fuel wood production.

Solar Energy

Solar energy endowments in all across the Commonwealth countries are substantially high. However, its extraction is still at a very low

TABLE 13.11 : Hydropower capability - Commonwealth

	Gross theoretical capability	Schemes >10MW		Exploitable capability for schemes
		Technically exploitable capability	Economically exploitable < 10 MW capability	
	GWh/yr	GWh/yr	GWh/yr	GWh/yr
AFRICA				
Ghana	12,782	>11,550	11,550	+
Kenya	>6,000	6,000	3,500	-
Lesotho	>2,000	2,000	-	+
Malawi	>6,000	6,000	-	-
Mauritius	150	110	-	-
Namibia	12,404	8,645	8,645	-
Nigeria	90,999	>2,278	2,278	-
Sierra Leone	>6,800	6,800	-	-
Swaziland	3,800	560	300	+
Tanzania	1,90,000	20,694	20,000	600
Zambia	>28,753	28,753	17,233	9
Zimbabwe	18,500	17,500	-	+
AMERICA, NORTH				
Canada	1,223,820	631,713	581,300	15,650
Dominica	200	136	130	+
Dominican Rep.	50,000	9,000	6,000	-
Jamaica	455	-	-	271
ASIA				
Cyprus	23,500	23,500	-	24
India (a)	2 637,800	600,000	600,000	75 000
Malaysia Pen.	123,000	12,290	-	-
Malaysia Sab/Sar.	107,000	>59,229	59,229	-
Pakistan	150,000	>120,000	120,000	-
Sri Lanka	10,000	>6,775	6,775	400
EUROPE				
United Kingdom	7,500	-	-	-
OCEANIA				
Australia	264,000	>30,000	30,000	1,248
New Zealand	152,082	77,200	43,910	4,280
Papua New Guinea	175,000	>98,000	98,000	+
Solomon Islands	37	37	37	-
Western Samoa	140	<80	80	+
Total	>5,302,722	>1,778,850	1,608,967	97,482
World Total	>34,230,264	>13,503,177,8	8,830,499	-
% of World Total	16	13	18	-

more than figures are taken as absolute figures.

Note: All data in Table refer to the Data Base of the World Energy Council and refers to the year 1993.

TABLE 13.12: Hydropower development for schemes >10MW - Commonwealth

	In operation		Under construction		Planned		Other
	Capacity	Actual gen. in 1993	Capacity	Probable annual gen.	Capacity	Probable annual gen.	annual gen.
	MW(e)	GWh	MW(e)	GWh	MW(e)	GWh	GWh
AFRICA							
Ghana	1,072	6,115	-	-	1,302	5,871	3,669
Kenya	604	2,993	0	0	60	335	-
Lesotho	-	-	70	500	110	-	-
Malawi	146	779	40	-	300	-	-
Mauritius	59	104	0	0	-	-	-
Namibia	249	1,080	0	0	450	1,650	-
Nigeria	1,938	5,252	-	-	3,550	28,542	-
Sierra Leone	-	-	222	-	310	-	-
Swaziland	43	130	-	-	15	60	-
Tanzania	328	1,697	246	1 043	2,226	11,170	-
Uganda	155	781	-	-	-	-	
Zambia	2,245	7,745	-	-	1,670	7,288	-
AMERICA, NORTH							
Canada	61,959	315,750	1 769	9 278	6,500	3,44,160	146,800
Dominican Rep.	375	1,294	72	-	300	-	-
ASIA							
Bangladesh	230	775	-	-	-	-	-
India	19,600	69,900	8 895	46 780	16,200	85,860	83,000
Malaysia Pen.	1,253	4,234	670	900	2,430	5,000	-
Malaysia Sab/Sar.	176	795	-	-	410	1,666	52,260
Pakistan	4,726	19,384	100	175	6,600	35,000	100,000
Sri Lanka	1,137	3,796	-	-	233	-	1,272
EUROPE							
United Kingdom	982	2,630	-	-	-	-	-
OCEANIA							
Australia	7,144	15,940	227	-	1,060	2,850	-
New Zealand	5,013	21,077	0	0	536	3,991	-
Papua New Guinea	218	457	-	-	107	-	-
Total	109,652	482,708	12,311	58,676	44,414	533,443	
World Total	609,264	22,85,960	-	-	-	-	
% of World Total	18	21	-	-	-	-	

Note: All data in Table refer to the Data Base of the World Energy Council and refers to the year 1993. -3 even negative.That is one of the options of reducing emissions in India.

TABLE 13.13: Hydropower development for schemes <10MW - Commonwealth

	In operation	Under construction and planned		
	Capacity MW(e)	Probable gen./yr GWh	Capacity MW(e)	Probable gen./yr GWh
AFRICA				
Kenya	7.4	-	-	-
Lesotho	2.2	-	0.2	-
Malawi	1	-	-	-
Mauritius	20	-	-	-
Namibia	0.1	-	-	-
Nigeria (<5)	32	-	1,100	-
Sierra Leone	4	25.66	2	-
Swaziland	8	-	0.3	2
Tanzania	11.4	-	60	-
Zambia (<1)	14	-	-	-
AMERICA, NORTH				
Canada	766	3,352	15	66
Dominica	14	37.8	-	-
Dominican Rep.	1.4	-	4.84	2
Jamaica	23.8	89.7	14.3	-
ASIA				
Cyprus	0.65	-	-	-
India (<15)	243	1,250	800	4,000
Malaysia Pen.	9.7	-	-	-
Malaysia Sab/Sr.(<2)	2.6	-	-	-
Pakistan	6.1	38.64	-	-
Sri Lanka	22.5	-	-	-
EUROPE				
United Kingdom	82	244	-	-
OCEANIA				
Australia	45	-	51	-
New Zealand	46	324	11	77
Papua New Guinea(<5)	22	-	-	-
Solomon Islands	0.03	0.07	8.24	45.5
Western Samoa	7.65	21.95	4	-
Total	1,392.53	5,383.82	2,070.88	4,192.5
World Total	34,166	72,129	-	-
% of World Total	4	8	-	-

Note: All data in Table refer to the Data Base of the World Energy Council and refers to the year 1993.

Group's activities should be widened to include certain aspects (the more efficient utilization) of conventional energy". In 1982, it was agreed that the scope of activities of the Group should be expanded to help promote co-operation in respect of feasibility studies, consultancy services, project execution etc for commercial scale energy projects in countries of the region.

Also, in 1982, Heads of Government specifically endorsed the proposals for further initiatives agreed at the third meeting of the Consultative Group (including the identification of 11 pilot projects and sites for the establishment of renewable energy systems and energy conservation technologies for the purpose of demonstration, evaluation and training), and welcomed the willingness of Australia and India to support the installation of these systems as part of the programme of activities of the Consultative Group. The Heads of Government, also recognizing that the work of the Consultative/Working Groups had entered a more action-oriented phase, took the view that the regional functional co-operation required a firmer basis for mobilizing regional financial resources. They agreed that additional earmarked contributions from member countries, on a voluntary basis, should, through a special account within the Commonwealth Fund for Technical Co-operation (CFTC), be used to finance activities undertaken at the request of the Groups on Agriculture, Energy, Trade and Maritime Issues.

The Industrial Development Unit (IDU) of the Commonwealth Fund for Technical Co-operation (CFTC), was designated as the executive agency for the pilot projects of the Group at the close of 1982. Thus, the Industrial Development Unit (IDU) assumed the role of assisting and facilitating the implementation of the various energy projects. Dr. Ranjit Watson, presently (1997) Chief Projects Officer at the Commonwealth Secretariate, was a dynamic force, at IDU, ensuring effective and timely implementation of projects.

IDU was actively ensuring that the pilot/demonstration projects of the Consultative Group were implemented in a manner that complemented and supplemented the other regional energy programmes including those co-ordinated by the South Pacific Economic Commission.

Furthermore, at the meetings of the Consultative Group, special efforts had been made to augment resources available to the Group by collaborating with the Asian Development Bank (ADB), Asia Pacific Development Centre (APDC), Asia Pacific Energy Planning Network (APENPLAN), ESCAP, SPEC and UN Pacific Energy Development Programme (UNPEDP).

TABLE 13.17: Energy Consumption

HDI high Human Development rank		Production as % of national energy reserves Coal 1991	Natural 1991	Crude 1991	Commercial Energy Production average annual growth rate% 1971-80	1980-92	Commercial Energy Consumption average annual growth rate (%) 1971-80	1980-82
1	Canada	0.8	3.4	11.0	3	4	4	2
11	Australia	0.2	4.6	18.0	5	6	3	2
17	New Zeland	2.3	4.3	8.2	5	8	3	5
18	United Kingdom	2.5	8.2	16.0	8	.	.	1
23	Cyprus	—	—	—	—	—	—	—
35	Singapore	—	—	—	—	—	8	7
39	Trinidad & Tobago	.	2.2	9.3	6	-(.)	4	4
41	Brunei	(.)	3.0	4.0	—	—	—	—
59	Malaysia	(.)	0.9	8.9	19	13	8	10
60	Mauritius	—	—	—	2	8	5	3
74	Botswana	—	—	—	9	(.)	11	3
88	Jamaica	—	—	—	(.)	-5	-2	3
97	Sri Lanka	—	—	—	8	8	2	1
126	Papua New Guinea	(.)	0	(.)	12	13	7	2
128	Pakistan	0.6	1.7	11.0	7	7	6	7
129	Ghana	—	—	—	7	2	3	2
130	Kenya	—	—	—	16	18	4	3
134	India	0.4	1.5	3.9	5	7	5	7
136	Zambia	—	—	—	7	-3	1	-3
141	Nigeria	(.)	0.2	3.9	3	2	19	1
146	Bangladesh	(.)	1.4	(.)	11	14	9	9
147	Tanzania	—	—	—	10	-1	2	-1
157	Malawi	—	—	—	11	4	8	1
158	Uganda	—	—	—	-4	2	-7	4
173	Sierre Leone	—	—	—	—	—	(.)	(.)

Over and above the delivery of programmes and projects on-the-ground, the Group had also been able to provide for an important dimension of technology exposure training, and exchange of experiences, whereby participants from member countries had been

TABLE 13.17: Energy Consumption

HDI Rank		Commercial energy use (kg of oil equivalent per capita)		Commercial energy efficiency (energy consumption equivalent per $ 100 GDP)		Commercial energy imports (as % of merchandise exports)	
		1971	1992	1965	1991	1971	1992
High Human development		587	1,241	126	39	—	11
1	Canada	6,261	7,912	279	50	5	4
11	Australia	4,035	5,263	167	30	4	6
17	New Zeland	2,448	3,743	121	39	8	7
18	United Kingdom	3,778	2,881	242	24	14	6
23	Cyprus	—	—	66	29	—	—
35	Singapore	1,551	4,399	124	43	23	15
39	Trinidad & Tobago	2,730	4,910	309	130	63	7
59	Malaysia	435	1,445	96	41	11	4
60	Mauritius	225	385	59	19	8	12
74	Botswana	247	395	162	15	—	—
88	Jamaica	996	1,075	131	59	23	28
97	Sri Lanka	81	101	59	37	2	12
126	Papua New Guinea	136	235	32	25	11	12
128	Pakistan	111	223	129	70	12	21
129	Ghana	106	96	54	31	8	52
130	Kenya	116	92	140	36	23	19
134	India	112	235	169	132	12	26
136	Zambia	335	158	129	80	7	21
141	Nigeria	40	128	30	45	1	1
146	Bangladesh	18	59	—	27	31	21
147	Tanzania	51	30	48	42	12	40
157	Malawi	37	40	35	18	17	28
158	Uganda	58	24	27	17	1	73
173	Sierre Leone	133	73	92	42	10	18

able to observe at first hand from their visits to sources of technology expertise involved in the development of energy projects in the region.

TABLE 13.18: Share of Commonwealth Countries in Commercial Sources of Energy
Percentage of World Total

		Resources	Production	Consumption
1.	Coal	17	14	12
		(24)(1)	(16)	(13)
2.	Oil	4	13	12
3.	Natural Gas	7	15	11
4.	Nuclear	—	9	—
5.	Hydro	16	21	—
		(2)	(3)	
6.	Fuelwood	(5)		

(1) Bituminous including authracite

(2) Gross Theoretical Capability > 10 MW

(3) Actual Generation > 10MW

(4) Probable Generation <10 MW

(5) Total Forest Area.

THE FUTURE COURSE

The future course for energy supply for upliftment of the poor in rural and urban areas should have a specific aim in the development planning of various Commonwealth countries. As the majority of the poor in these countries are residing in the rural areas, an effort to concentrate on the uniqueness of the rural energy planning and implementation will be urgently required so as to achieve the best possible results. Having been involved with the Commonwealth Project for Energy Supply in the Asia-Pacific, as detailed earlier, a deeper understanding of the site specific conditions has helped me to identify the approach. Following details can form an indicative method for improving energy supply for the millions in Commonwealth Countries.

Rural Energy Supply Options

Realising that the conventional energy supply systems that have evolved in conjunction with large scale, highly concentrated production and distribution units have often proved to be inadequate for the needs of the rural population in Commonwealth developing countries. They have often failed to yield adequate electrical power in relation to the micro-economic needs; and have usually not met the total specific demands of agriculture or of rural and small scale industries.

When properly used, small and perhaps intermittent amounts of

TABLE 13.19: Human Development Index

HDI Rank		Life expectancy at birth (Years) 1992	Adult Literacy rate (%) 1992	Real GDP Per Capita per Capita (PPP$) 1992
High Human development		72.9	95.8	13,605
1	Canada	77.4	99.0	20.520
11	Australia	77.6	99.0	18.220
17	New Zealand	75.5	99.0	14.990
18	United Kingdom	76.2	99.0	17.160
23	Cyprus	77.0	94.0	15.050
25	Barbodos	75.6	97.0	9.667
29	Belize	73.6	96.0	5.619[b]
34	Malta	76.1	87.0	8.281[b]
35	Singapore	74.8	89.9	18.330
37	St Kits and Nevis	70.0	99.0	5.177
39	Trinidad and Tobago	71.6	97.4	9.760
41	Brunei Darussalam	74.2	86.4	20.589[b]
55	Antigua and Barbuda	74.0	96.0	4.436[b]
59	Malaysia	70.8	81.5	7.790
60	Mauritius	70.2	81.1	11.700
61	Seychelles	71.0	77.0	5.619[b]
67	Grenada	70..0	98.0	3.822[b]
69	Dominica	72.0	97.0	3.526[b]
74	Botswana	64.9	67.2	5.120
79	Saint Vincent	71.0	98.0	3.322[b]
84	Saint Lucia	72.0	93.0	3.026[b]
88	Jamaica	73.6	83.7	3.200
97	Sri Lanka	71.9	89.3	2.850
102	Samoa (Western)	67.6	98.0	1.869
105	Guyana	65.2	97.5	1.800
108	Namibia	58.8	40.0	4.020
118	Maldives	62.1	92.6	1.200
119	Vanatu	65.2	65.0	1.956[b]
124	Swaziland	57.5	74.0	1.700
125	Solomon Islands	70.4	24.0	2.616[b]
126	Papua New Guinea	55.8	69.7	2.410
128	Pakistan	61.5	35.7	2.390
129	Ghana	56.0	60.7	2.110
130	Kenya	55.7	74.5	1.400
131	Lesotho	60.5	68.6	1.060
134	India	60.4	49.9	1.230
136	Zambia	48.9	75.2	1.230
141	Nigeria	50.4	52.5	1.560
146	Bangladcsh	55.6	36.4	1.230
157	Malawi	45.6	53.9	820
158	Uganda	44.9	58.6	860
161	Gambia	45.0	35.6	1.260
173	Sierra Leone	39.0	28.7	880

b. Preliminary update of the Penn World Tables using an expanded set of international comparisons, as described in Summers and Heston 1991.

Ref: UNDP Human Development Report 1995.

TABLE 13.14: Forest area and fuelwood (including charcoal) production-Commonwealth

	Total forest area 10^6 h	Productive forest area 10^6 ha	Fuelwood production ktonnes
AFRICA			
Botswana	26.6	14.3	979
Gambia	0.3	0.1	679
Ghana	18	9.6	11,246
Kenya	16.8	1.3	26,615
Lesotho	-	-	472
Malawi	3.7	3.6	6,921
Mauritius	-	-	1
Nigeria	65.7	15.8	79,597
Sierra Leone	7	1.9	2,311
Swaziland (a)	0.1	0.1	406
Tanzania	68.5	33.7	24,623
Uganda	16	6.4	9,788
Zambia	60.3	32.3	9,390
AMERICA, NORTH			
Canada (R)	416	237	7,528
Dominican Rep.	1.5	1.1	708
Jamaica (R,a)	0.3	-	7
Trinidad & Tobago	0.2	0.2	16
ASIA			
Bangladesh	1.5	1	23,036
Brunei	0.5	0.5	57
Cyprus	0.3	0.1	13
India (a)	82.6	65	1,90,517
Malaysia	22.2	17.7	6,797
Pakistan	3.1	2	18,140
Singapore	-	-	87
Sri Lanka	4	1.9	6,310
EUROPE			
United Kingdom	2.4	2.2	185
OCEANIA			
Australia	146	39.8	2,101
New Zealand (R,a)	7.7	1.5	36
Papua New Guinea	42.1	36	4,011
Solomon Islands	2.5	2.4	100
Vanatu	0.8	0.8	17
Total	1,016.7	564.3	432,694
World Total	4,968	3,249	1,366,529
% of World Total	21	17	32

Note: All data in Table refer to the Data Base of the World Energy Council and refers to the year 1993.

Range of Project Activities

This Group had undertaken the difficult task of translating several outcomes of research and development in the non-conventional and renewable energy sector, into practical projects for implementation. Many of the projects were located in the small island countries of the Pacific and had to be specially designed for site specific conditions of implementation and of technology adaptation and technology transfer which required close attention of co-operative sources in the region as several technologies were then at the pre-commercialization stage.

Projects completed were relatable to priorities in foreign exchange savings and to adapting renewable and non-conventional energy technologies for use in isolated areas as well as in rural areas of the region. Projects completed range over:

Project Range

(i) Energy audits for energy conservation and management in commercial and industrial establishments.
(ii) Solar thermal and photovoltaic applications for small power users;
(iii) Solar Crop Drying;

Spread

Some of the projects may be mentioned to give an idea of the spread:

(i) Biomass Technology:
— natural gas fuel substitution for urban transport in Bangladesh.
— producer gas unit reactivation and development for institutional electric power supply in Vanuatu.
— Retrofitting of boiler and oven installations with biomass gasifiers/hot air generators in Western Samoa.

(ii) Energy Audit:
— Implementation of energy audit recommendations in hospitals in Fiji, Solomon Islands, Tonga and Vanuatu.

(iii) Solar Thermal and Photovoltaic Applications:
— development of solar and wind energy sources for irrigation in Tonga.
— photovoltaic packages for village use in the Maldives and Western Samoa.

TABLE 13.15: National wind energy targets - Commonwealth

Country	Target in MW(e)
Canada	30
Greece	550
India	3 000
UK	740(2)

Note: All data in Table refer to the Data Base of the World Energy Council and refers to the year 1993.

TABLE 13.16: Installed capacity and annual output for electricity generation from wind energy (data refer to 1993 unless otherwise noted) - Commonwealth

	Installed capacity MW_e	Annual output GWh
Argentina	0.12	0.32
Australia	0.2	0.43
Canada (a)	1.05	2
Greece (a)	6	45
India (a)	40	45
New Zealand	0.23	1
United Kingdom (a)	130	211
Total	197.6	304.75
World Total	3 517.37	4 877.04
% of World Total 6		6

Note: All data in Table refer to the Data Base of the World Energy Council and refers to the year 1993.

— Solar water heating for hospital in Fiji.

(iv) Solar Crop Drying:

— cassava, ginger, pepper and vanila in Tonga.

— copra in Solomon Islands.

(v) Waste Heat Recovery:

— Waste heat recovery systems for commercial and industrial applications in Fiji and Papua New Guinea.

Impact of the Activities and Projects

The progress of projects and its further directions were reviewed at Technical Workshops held from time to time. Criteria for success were mentioned as follows:

(i) co-ordination with other energy programmes in the region to avoid duplication;

(ii) extensive consultations to properly identify needs and formulate a realistic programme;

(iii) criteria for selection of proposed projects which recognized that:

— the technologies should be appropriate for the location under consideration

— technologies which have not been proven to be technically feasible are unlikely to be of value to developing countries.

— any demonstration projects should be supported by full technical follow through, i.e., monitoring/assessment/dissemination of results and any necessary training, etc.

— consideration should be given to any social issues that might arise (e.g. the impact of increased demand for wood for a wood-fired power plant).

(iv) prompt initiation and implementation of projects following their formulation using well defined procedures for the appointment of consultants, purchase of equipment, etc in order to ensure that the initial enthusiasm is maintained and the objectives remain relevant;

(v) flexible administration to accommodate changing circumstances or unavoidable changes such as long delays, frustrations and loss of credibility which can be avoided if decisions on all but major changes can be made without the need to refer back to a policy making body;

(vi) full technical support throughout the life of the project through regular monitoring and assessment,and any necessary training which will enable potential problems to be identified and avoided at an early stage;

(viii) prompt compilation of a final report.

The Group had been able to;

(i) catalyse regional co-operation in terms of technology transfer for project implementation from identified sources in the developing and developed countries of the region;

(ii) exchange information on technologies and projects for facilitating further project implementation and for ensuring that counterpart capabilities are developed in recipient countries for operation and maintenance of projects; and

(iii) mobilise financial and other resources for implementation of projects from commercial sources, such as the Asian Development Bank.

TABLE 13.19: Human Development Index

HDI Rank		Life expectancy at birth (Years) 1992	Adult Literacy rate (%) 1992	Real GDP Per Capita per Capita (PPP$) 1992
High Human development		72.9	95.8	13,605
1	Canada	77.4	99.0	20.520
11	Australia	77.6	99.0	18.220
17	New Zealand	75.5	99.0	14.990
18	United Kingdom	76.2	99.0	17.160
23	Cyprus	77.0	94.0	15.050
25	Barbodos	75.6	97.0	9.667
29	Belize	73.6	96.0	5.619[b]
34	Malta	76.1	87.0	8.281[b]
35	Singapore	74.8	89.9	18.330
37	St Kits and Nevis	70.0	99.0	5.177
39	Trinidad and Tobago	71.6	97.4	9.760
41	Brunei Darussalam	74.2	86.4	20.589[b]
55	Antigua and Barbuda	74.0	96.0	4.436[b]
59	Malaysia	70.8	81.5	7.790
60	Mauritius	70.2	81.1	11.700
61	Seychelles	71.0	77.0	5.619[b]
67	Grenada	70..0	98.0	3.822[b]
69	Dominica	72.0	97.0	3.526[b]
74	Botswana	64.9	67.2	5.120
79	Saint Vincent	71.0	98.0	3.322[b]
84	Saint Lucia	72.0	93.0	3.026[b]
88	Jamaica	73.6	83.7	3.200
97	Sri Lanka	71.9	89.3	2.850
102	Samoa (Western)	67.6	98.0	1.869
105	Guyana	65.2	97.5	1.800
108	Namibia	58.8	40.0	4.020
118	Maldives	62.1	92.6	1.200
119	Vanatu	65.2	65.0	1.956[b]
124	Swaziland	57.5	74.0	1.700
125	Solomon Islands	70.4	24.0	2.616[b]
126	Papua New Guinea	55.8	69.7	2.410
128	Pakistan	61.5	35.7	2.390
129	Ghana	56.0	60.7	2.110
130	Kenya	55.7	74.5	1.400
131	Lesotho	60.5	68.6	1.060
134	India	60.4	49.9	1.230
136	Zambia	48.9	75.2	1.230
141	Nigeria	50.4	52.5	1.560
146	Bangladesh	55.6	36.4	1.230
157	Malawi	45.6	53.9	820
158	Uganda	44.9	58.6	860
161	Gambia	45.0	35.6	1.260
173	Sierra Leone	39.0	28.7	880

b. Preliminary update of the Penn World Tables using an expanded set of international comparisons, as described in Summers and Heston 1991.

Ref: UNDP Human Development Report 1995.

TABLE 13.18: Share of Commonwealth Countries in Commercial Sources of Energy
Percentage of World Total

		Resources	Production	Consumption
1.	Coal	17	14	12
		(24)(1)	(16)	(13)
2.	Oil	4	13	12
3.	Natural Gas	7	15	11
4.	Nuclear	—	9	—
5.	Hydro	16	21	—
		(2)	(3)	
6.	Fuelwood	(5)		

(1) Bituminous including authracite

(2) Gross Theoretical Capability > 10 MW

(3) Actual Generation > 10MW

(4) Probable Generation <10 MW

(5) Total Forest Area.

THE FUTURE COURSE

The future course for energy supply for upliftment of the poor in rural and urban areas should have a specific aim in the development planning of various Commonwealth countries. As the majority of the poor in these countries are residing in the rural areas, an effort to concentrate on the uniqueness of the rural energy planning and implementation will be urgently required so as to achieve the best possible results. Having been involved with the Commonwealth Project for Energy Supply in the Asia-Pacific, as detailed earlier, a deeper understanding of the site specific conditions has helped me to identify the approach. Following details can form an indicative method for improving energy supply for the millions in Commonwealth Countries.

Rural Energy Supply Options

Realising that the conventional energy supply systems that have evolved in conjunction with large scale, highly concentrated production and distribution units have often proved to be inadequate for the needs of the rural population in Commonwealth developing countries. They have often failed to yield adequate electrical power in relation to the micro-economic needs; and have usually not met the total specific demands of agriculture or of rural and small scale industries.

When properly used, small and perhaps intermittent amounts of

TABLE 13.17: Energy Consumption

		Commercial energy use (kg of oil equivalent per capita)		Commercial energy efficiency (energy consumption equivalent per $ 100 GDP)		Commercial energy imports (as % of merchandise exports)	
HDI Rank		1971	1992	1965	1991	1971	1992
High Human development		587	1,241	126	39	—	11
1	Canada	6,261	7,912	279	50	5	4
11	Australia	4,035	5,263	167	30	4	6
17	New Zeland	2,448	3,743	121	39	8	7
18	United Kingdom	3,778	2,881	242	24	14	6
23	Cyprus	—	—	66	29	—	—
35	Singapore	1,551	4,399	124	43	23	15
39	Trinidad & Tobago	2,730	4,910	309	130	63	7
59	Malaysia	435	1,445	96	41	11	4
60	Mauritius	225	385	59	19	8	12
74	Botswana	247	395	162	15	—	—
88	Jamaica	996	1,075	131	59	23	28
97	Sri Lanka	81	101	59	37	2	12
126	Papua New Guinea	136	235	32	25	11	12
128	Pakistan	111	223	129	70	12	21
129	Ghana	106	96	54	31	8	52
130	Kenya	116	92	140	36	23	19
134	India	112	235	169	132	12	26
136	Zambia	335	158	129	80	7	21
141	Nigeria	40	128	30	45	1	1
146	Bangladesh	18	59	—	27	31	21
147	Tanzania	51	30	48	42	12	40
157	Malawi	37	40	35	18	17	28
158	Uganda	58	24	27	17	1	73
173	Sierre Leone	133	73	92	42	10	18

able to observe at first hand from their visits to sources of technology expertise involved in the development of energy projects in the region.

TABLE 13.17: Energy Consumption

HDI high Human Development rank		Production as % of national energy reserves Coal 1991	Natural 1991	Crude 1991	Commercial Energy Production average annual growth rate% 1971-80	1980-92	Commercial Energy Consumption average annual growth rate (%) 1971-80	1980-82
1	Canada	0.8	3.4	11.0	3	4	4	2
11	Australia	0.2	4.6	18.0	5	6	3	2
17	New Zeland	2.3	4.3	8.2	5	8	3	5
18	United Kingdom	2.5	8.2	16.0	8	.	.	1
23	Cyprus	—	—	—	—	—	—	—
35	Singapore	—	—	—	—	—	8	7
39	Trinidad & Tobago	.	2.2	9.3	6	-(.)	4	4
41	Brunei	(.)	3.0	4.0	—	—	—	—
59	Malaysia	(.)	0.9	8.9	19	13	8	10
60	Mauritius	—	—	—	2	8	5	3
74	Botswana	—	—	—	9	(.)	11	3
88	Jamaica	—	—	—	(.)	-5	-2	3
97	Sri Lanka	—	—	—	8	8	2	1
126	Papua New Guinea	(.)	0	(.)	12	13	7	2
128	Pakistan	0.6	1.7	11.0	7	7	6	7
129	Ghana	—	—	—	7	2	3	2
130	Kenya	—	—	—	16	18	4	3
134	India	0.4	1.5	3.9	5	7	5	7
136	Zambia	—	—	—	7	-3	1	-3
141	Nigeria	(.)	0.2	3.9	3	2	19	1
146	Bangladesh	(.)	1.4	(.)	11	14	9	9
147	Tanzania	—	—	—	10	-1	2	-1
157	Malawi	—	—	—	11	4	8	1
158	Uganda	—	—	—	-4	2	-7	4
173	Sierre Leone	—	—	—	—	—	(.)	(.)

Over and above the delivery of programmes and projects on-the-ground, the Group had also been able to provide for an important dimension of technology exposure training, and exchange of experiences, whereby participants from member countries had been project the South Asia position in proper context. The forerunner to this Report was the WEC Commission's Report titled "Energy for Tomorrow's World" giving projections upto 2020. The studies have indicated that 90 per cent of the increase in demand could take place in the currently defined developing countries, particularly in Asia, and the Latin America. It means that the developing countries which are today consuming 30 per cent of the world's total energy will consume 50 per cent by 2020. The balance will tilt further towards the developing countries.[2]

It does not mean that the development process in the developed countries will slow down; but it simply shows that development process in the developing countries will increase at a pace never witnessed in the past. At the same time, since most of it is expected from inefficient or low efficiency technologies, the rate of consumption of primary energy for the same final output will be higher in the developing countries. Low capacity of the developing countries to finance the emerging energy technologies will also be a contributory factor.

ASIA ENERGY VISION 2020 - AN EFFORT TO IDENTIFY THE OPTION FOR SUSTAINABLE ENERGY SUPPLY

The International Conference on Energy: Asia Energy Vision 2020 organised under the auspices of the Indian Member Committee—World Energy Council identified nine basic steps that will fulfil the energy needs in the South Asia region. These cover broadly the supply options, the pollution problem, and the financing of the energy projects. The role of regional and international cooperation in achieving these objectives are considered important, both on government level as well as the industry level.[5] In brief, the recommendations are as follows:

(i) To tackle rural poverty and energy deprivation through the installation of decentralised, locally appropriate, and affordable renewable energy systems.

(ii) To raise the efficiency of energy production and use at every point—from modern, centralised power generation systems to end-uses—and to provide the necessary support for more environmentally attractive options.

(iii) To address, much more vigorously than at present, the problem of local and regional pollution—from the use of fuelwood through to cleaner fossil fuel production and reduced emissions from road transport.

(iv) To build up the necessary infrastructure and attract the required investment and its financing—which requires, in turn, realism about economic decision-making and how investible funds are successfully mobilised—domestically and internationally.

(v) To phase out subsidies and eventually to price energy at a level which covers all costs, while recognising the inevitable realities for the poor of inability to pay, which requires innovative financing schemes,

(vi) To liberalise energy markets and privatise energy institutions in order that efficiency, availability and investment can be increased and the benefits of competition and market signals realised.

(vii) To raise the level and effectiveness of Research, Development and Demonstration expenditures.

(viii) To reinforce the effectiveness of international cooperation, particularly at the regional level.

(ix) To require the industrialised countries to fulfil their agreed commitments to contribute towards technology and financial transfers, to meet their agreed environmental aims, and to demonstrate their wider obligations to humanity.

It was identified that the policy makers, as their prime duty, had to create the necessary conditions for social development and economic output, while conserving environmental features. This is the essence of sustainable development.

The Conference was recognised as a major step in promoting international cooperation, mainly at the regional level. The Conference also helped in better understanding amongst regional participants.

The New Delhi Declaration giving details of Conclusions and Recommendations is given at ANNEXURE.

GLOBAL THEMES UNDERGOING REVOLUTION

The Conference also realised that the three global themes undergoing revolution in the continuing globalisation of energy markets are as follows:[4]

(i) Energy markets today are being served by global companies in just not oil but in coal,electricity and gas also;

(ii) Governments are increasingly looking outside their own ambit for resources to develop national energy systems;

(iii) Technology and investments are flowing to those energy markets that are giving expected returns to the multinational companies, rather than those that fulfil the needs of the recipient countries.

IMPORTANT FACTORS HAVING IMPACT ON FUTURE ENERGY SUPPLY

Three important factors which will interact with each other for the future energy supply were identified as follows:[5]

(i) technical progress;
(ii) demographic evolution of the world populations;
(iii) their standards of living and a heightened perception of the need to respect the conditions of life on the planet.

The effect of *technical progress* over the next 25 years may not be so easy to predict accurately, but certain visionary possibilities can be projected. An example is relevant for discussion at PETROTECH 97:

The delivered price of gas reflected for the most part the cost of its transport. The progress in this direction is not expected to be as spectacular as advances in petroleum technology and gas transport will continue to represent a major disadvantage. Natural gas will benefit from a certain advantage because of the greenhouse effect. In gas liquefaction, the cost of fixed infrastructure is such that more flexibility in transport can be expected because attempts will be made to fully utilise the available capacities. A real liquefied natural gas market similar to that of oil is likely to emerge.

With the rise in world energy demand brought about by the *population explosion* and other factors, we will be confronted by the fundamental limitation of energy, namely, the finite nature of existing usable energy resources. It will be important at that time that we overcome the problems created by the uneven distribution of fossil fuels by establishing international coordination and ties of international inter-dependence. This will have to be backed up by promoting energy conservation measures and securing a stable energy supply through technological innovation and R&D, in order to overcome the problems resulting from the finite nature and unbalanced distribution of energy sources.

SIGNIFICANCE OF INTERNATIONAL COOPERATION

Asia Energy Vision 2020 projected following important issues for international cooperation in the energy sector:

i. Interdependence of Nations

The future will also see an *interdependence of nations* owning primary energy sources and those needing it. While it is easy for the developed industrialised nations to establish this linkage, by promoting economic cooperation and technology transfer, basically through open market economy and private sector intervention, the interaction between the developing countries and the energy rich countries will have to come through bilateral trade agreements and only through governments' intervention. The option of private sector led market intervention have their own potential and must be vigorously pursued.

ii. Pace of International Energy Cooperation

The pace of international energy cooperation and coordination was observed to have been slow, and given the magnitude of the problem likely to confront the energy and energy related sectors over the next 30 years it was considered as a major issue requiring considerable attention and development. This issue cannot be left only to the market forces to be decided in the industrialised countries by the multinational corporations. These are the issues that have to be dealt with by the government agencies and inter-governmental agencies through continuous dialogue and negotiation.

iii. Widest Possible Range of Energy Options to be kept open

The widest possible range of economic and publically acceptable options are required to be kept open—including accelerating renewable energy development and maintaining the nuclear option, as well as recognising fossil fuel dependency. That strategy needs to be implemented without further delay. The necessary actions were called to start now. The lead times to achieve major change in the energy consumption patterns are long. Therefore, the policies need to be expressed with clear time tables and appropriate actions. The appropriate actions will recognise current needs and future demands, local and regional priorities, and what support can realistically be expected from elsewhere—from the industrialised countries, international organisations and multinational energy companies.

iv. Innovation in International Financing

Innovation in international financing was considered necessary. At present, the developing countries—that includes India—are looking out for investments both in the infrastructure sector as well as in the generating sector. To ensure that infrastructure is in place as per the government's requirements, stricter bureaucratic procedures have to be followed as there is an accountability to the Parliament and to the Office of the Comptroller and Auditor General for funds going out of the government's exchequer. At the same time, the actions and aspirations of the millions in India have to be fulfilled, and therefore, fuel pricing policy cannot be replaced by an overall market driven energy pricing as is being practiced in the industrialised world. It obviously needs innovative financing mechanism towards which the industrialised world and the financing agencies and institutions have to look into and evolve an appropriate mechanism (that allows for accommodation of subsidies as an inseparable part during the transitory phase). Whereas the private sector, both investors and financial institutions will seek privileges and protection before allocating large resources in the form of capital and loans, the government in turn will accept the private sector to perform socially and economically desirable duties and obligations. What is really being suggested under the new financing mechanism by the developed world is a series of bureaucratic and procedural changes in the government's functioning.

The investment in green fuel sources will obviously be higher and require longer term lock-in periods. Natural gas is expected to play an important role to year 2050 when the renewables may contribute significantly higher. Such a transition also calls for innovation in international financing. The initial start-up period of the industry has to be supported by a regulatory framework that gives the investor the confidence that his investments are totally protected. No one will put huge sums of money in pipelines and then provide open access to them, so that the competitors can get a profit, while they bypass a risk.

v. International Cooperation in Manpower for Transfer of Technology

Another area for international cooperation is manpower for transfer of technology. Many a time the recipient governments feel that the projects involving technology transfer could equally be accomplished by local experts. Local knowledge indeed is essential in any project, and should be strengthened representing any joint ventures and other strategic alliances. However, when it comes to realising a large and complex project, the international energy company can bring discipline as is relevant to large enterprises. Technical knowhow and management capability will ensure proper management, besides possible implementation. Quite simply because the international operator will have done it before. The international energy company may also transfer to the host country advanced application technologies which will enable indigenous industries, such as cement or ceramic, to benefit from gas use know-how.

vi. Keeping Energy Supply Independent of Political Considerations

ESCAP projected a strategy to keep energy supply independent of political considerations thereby the share of energy supplied by natural gas, hydro power and renewable energy can be increased in countries of South Asia.[5]

The most important is the natural gas supply. Whereas the local supplies for Bangladesh, India and Pakistan are predicted at 30 to 40 years of their present consumption levels, it is also recommended that a parallel approach for India and Pakistan is to import the natural gas from the Middle East or Central Asia where several of the countries have very large reserves to production ratios. Several options for importing natural gas have already been examined in India and Pakistan. India has been exploring the possibility of importing natural gas via pipeline or in liquefied form from Oman or via pipeline from the Islamic Republic of Iran. Pakistan has signed Memoranda of Understanding with Islamic Republic of Iran, Qatar, Turkministan, for the possible supply of natural gas. Although India and Pakistan could make independent arrangements for importing natural gas, economies of the countries suggest that each of the countries would benefit a great deal by cooperating in the construction of one or more pipelines that could supply natural gas to both countries. This can only be achieved by regional cooperation as each of the countries may find costs very high to meet individual requirements.

Such a projection has been made on the basis that economic and political necessities can be separated and dealt on their own merits.

CONCLUSION

Long term energy sustainability calls for innovative global approach promoting new global cooperation amongst nations on bilateral and multilateral approaches. Whereas open market economy will regulate energy supply in the industrialised world, government interventions in developing countries will be important to ensure long term sustainable energy supplies.

International cooperation is also crucial for technical and financial transfers to developing countries to build capabilities in the energy sector. Interdependence will be the key to sustainable energy supply.

REFERENCES

1. Global Energy Perspectives to 2050 and Beyond - WEC.
2. A Global Energy Overview - Ian D. Lindsay (WEC) - Asia Energy Vision 2020.
3. New Delhi Declaration - Asia Energy Vision 2020.
4. Future Energy Markets - John Baker (WEC) - Asia Energy Vision 2020.
5. South Asian Cooperation for Enhancing Energy Supply in the 21st Century - Toufiq A. Siddiqui (ESCAP) - Asia Energy Vision 2020.
6. New Delhi Declaration - Asia Energy Vision 2020.

ANNEXURE

NEW DELHI DECLARATION - 1996

Sustainable Energy Supply and Asian Development

This New Delhi Declaration forms the Conclusions and Recommendations of the International Conference: Asia Energy Vision-2020, held in New Delhi 15-17 November, 1996. The Conference fully confirmed the World Energy Council's focus in its "Message for 1997" on the plight of 2 billion people in the world now without adequate energy—800 million of them in the South Asia Region—and the challenge of meeting future requirements in a world where population is likely to double over the next century.

The challenges of inadequate water and food supplies are closely linked to those of reliance on fuelwood and agro wastes. Failure to meet them results in comprehensive deprivation. More than half these deprived live in the South Asia Region.

Policy makers have, as their prime duty, to create the necessary conditions for social development and economic output while conserving essential environmental features. This is the essence of sustainable development. For sustainable development to occur, a clear strategy is required. The widest possible range of economic and publicly acceptable options needs to be kept open—including accelerating renewable energy development and maintaining the nuclear option, as well as recognising fossil fuel dependency. That strategy needs to be implemented without further delay. The necessary actions must start now. The lead times to achieve major change are long. Policies need to be expressed with clear timetables and action programmes. The appropriate actions will recognise current needs and realities, local and regional priorities, and what support can realistically be expected from elsewhere—from the industrialised countries and international organisations.

Rural poverty, and the need to create conditions for rural economic and social development, is the first need to be addressed. The necessary response is the installation of decentralised, locally appropriate and affordable renewable energy systems. This must be the *prime goal* in South Asia.

However energy is supplied, it should be conserved and used efficiently. Where energy is imported, the impacts on the balance of payments can be greatly reduced. The scope for improving energy

efficiency is huge. Without efficiency improvements the supply of energy—whatever its source—is more difficult and costly. It is generally cheaper to invest in raising energy efficiency than in adding to capacity. For providing energy to large centres of population and large industrial complexes—which need to compete domestically and internationally—efficient and environmentally acceptable centralised systems of power generation and distribution are required. The *second goal* for energy policy in South Asia is to raise energy efficiency.

Pollution is a severe problem in South Asia. Energy provision and use is a major source of this pollution. In homes using fuelwood, agro wastes, and inefficient cooking stoves air quality is often foul and respiratory diseases and death an all too frequent result. The use of coal without adequate application of clean coal technology is another important source of pollution. The road transport sector is another, rapidly growing, source of pollution. The *third goal* is to address, much more vigorously than at present, the problems of local and regional pollution in South Asia by a combination of improved efficiency and better technology.

Many of the energy supply problems in South Asia - which inevitably are reflected in demand side issues - flow from lack of infrastructure. The building up of the necessary infrastructure requires investment, which in turn means financing and with adequate rewards going to investors. This can only be achieved by a combination of mobilising domestic savings, supported by the necessary local institutions, and attracting inward foreign investment on acceptable terms. There are many competing opportunities for international investment. South Asia needs to create the conditions which make the Region a relatively attractive area to invest in. This means both adequate returns to foreign and domestic investors, and avoidance of undue procedural obstacles, and institutional barriers. The *fourth goal* for South Asia, therefore, is to strengthen the necessary infrastructure and attract the investment and its financing. This will require realism about economic decision-making and creating the conditions for attracting foreign investment. Improved public education and awareness will raise understanding of economic realities and receptivity to technological innovation.

Much of what has been stated in this Declaration means trying to kick-start sustainable energy development when, on the one hand, huge numbers of people in South Asia cannot afford commercial energy, while on the other hand, they will not obtain commercial energy unless they pay a price which covers the costs of provision. For as long as the electricity tariff in rural India is little more than 3 per cent of total costs of supply, and little more than 25 per cent of fuel costs alone, supplies will remain severely inadequate; the necessary infrastructural investments will not take place; and the technology and finance desired from other countries and international organisations will not be forthcoming. Innovative financing mechanisms will be required in many cases to help bridge the gap between the present situation and the introduction of full market prices. The *fifth goal* for South Asia, therefore, is to phase out energy subsidies and eventually to move to pricing energy at a level which covers all costs—including environmental costs.

Institutional changes are required to allow competition to work more effectively and efficiency to be increased. Liberalisation of policy and privatisation are required, with the retention of the necessary framework to ensure effective competition, conformation to legislative requirements, and the setting of tariffs which are mutually acceptable. There is a growing number of examples in South Asia of how industrial performance can be improved by new, commercial management—reflecting the manifold professional skills of the people of South Asia. These professional skills include engineering, financial, economic, and entrepreneurial skills as well as those of modern management. The availability of these kills is one of the necessary conditions for the future energy development of the Region. The *sixth goal* for the South Asia Region, therefore, is the liberalisation and privatisation of the whole energy supply sector in order that efficiency, availability, and investment can be improved.

Technological innovation and diffusion, together with supportive institutional changes, will have a powerful role to play in promoting sustainable energy provision and use. Underlying these developments will need to be effective Research, Development and Demonstration expenditures. The *seventh goal* for the South Asia region is, therefore, to raise the level and effectiveness RD&D expenditure.

Many of the challenges outlined above will be more successfully addressed within the context of international, including regional co-operation. The *eighth goal* for South Asia is to reinforce the effectiveness of regional co-operation in particular.

Finally, the New Delhi Declaration notices that the industrialised countries have failed, and continue to fail, to fulfil the international commitments they have freely entered into to provide financing and technology transfers, and to meet the environmental goals they aim at and the wider responsibilities they acknowledge. The *ninth goal* for the South Asia Region is to call successfully upon the industrialised countries to meet their agreed obligations with respect to the Region.

SUGGESTED FOLLOW-UP ACTION

Following Five Activities have been suggested:

1. Regional Workshop on Biomass as a source of sustainable energy supply.
2. Regional Workshop on mini-hydro system.
3. Regional Workshop on Energy Conservation and Efficiency.
4. Regional Workshop on Nuclear Power.
5. National Workshop on R&D Needs in the Energy Sector in India.

15

NEED FOR COOPERATION IN ASIA PACIFIC REGION

INTRODUCTION

Malaysia has progressed significantly through appropriate policies, plans and timely and effective implementation strategies. A sustained growth of GDP at over 7 per cent for 8 years has resulted in social development of the masses through sound economic development. This was also appreciated at the World Summit for Social Development held at Copenhagen during March 6-12, 1995. We are meeting so close to the Summit that all strategies for energy availability and supply shall have to be considered in the light of the development decisions taken by the Head of States.

The Copenhagen Declaration adopted by the Heads of Government mentions 'We are deeply convinced that economic development, social development and environmental protection are interdependent and mutually reinforcing components of sustainable development, which is the framework for our efforts to achieve a higher quality of life for all people. Equitable social development that recognises empowering of the poor to utilise environmental resources sustainably is a necessary foundation for sustainable development. We also recognise that broad-based and sustained economic growth in the context of sustainable development is necessary to sustain social development.

Beyond appreciation for the efforts of Malaysia, the World Summit for Social Development also focussed on the poverty and

---

Paper presented at the International Workshop on Energy organised by the University Kabangsaan, Malaysia as a part of 25th Year Celebration at Kabangsaan during May 17-20, 1995.

unemployment problems of the Asia Pacific Region. Therefore, any discussion at this conference when viewed in the overall regional context will project the possibilities of local and regional solutions that will be community based and utilising the local resources. More so that will also be responsive to the needs of the masses. Availability of energy will be an important component of the whole approach.

SITUATION IN INDIA

India is basically a rural economy with 75 per cent of its population still living in the rural areas. Now there is a tendency for fast pace of urbanisation and creation of mega cities leading to the associated urban problems. The changes are quite perceptible and the percentage of population in the urban centres is growing. The numbers are increasing at an enormous rate, both in the rural as well as urban centres. On the other hand sudden and unplanned growth of population stresses the energy supply system.

The overall demand growth in the rural sectors is comparatively low for reasons of:

(i) migration of populations to the urban centres; and
(ii) larger areas of rural sector becoming urbanised.

On the other hand, the growth of energy requirement in the urban centres is a major concern of all the mega cities in the country, e.g. the capital city of Delhi has a shortfall of over 600 MW installed capacity for electricity supply during the peak load time in the summer months this year.

Demand for energy in the rural household sector was traditionally met through non-commercial sources, namely, firewood, crop residues, animal waste, as well as human and drought animal power. The levels of efficiency at which useful energy is being realised from these sources are very low—varying between 10 to 15 per cent. Possibilities of improving efficiencies of use have been demonstrated not only on the lab scale but also on the field level. However, non-commercial energy sources are steadily getting replaced by other commercial sources of energy which provide energy of much higher quality and efficiency and involve at the same time high capital cost.

POLICY FOR PROMOTION OF FOREIGN AND PRIVATE INVESTMENT IN INDIA

Power Sector

The major problem in creation of power generation capacity in India is the non-availability of institutional finance and not the non-availability of suitable energy resources. Capital cost of thermal power plants is Rs. 25 to 35 million per MW. Foreign Investment Projects in the power sector (e.g. ENRON) also project the capital cost to be as high as Rs. 40 million per MW, and generation cost of Rs. 2.39 per unit (8 cents US/Kwh).

Realising the need for speedy increase in the energy sector, the government decided on specific measures to promote foreign financial investments. Some specific measures were taken in 1992-93 to bring about structural reforms in the power sector. The Policy Package for inviting private investments included the following:

(i) Reduction of import duties on power equipment to 20 per cent.
(ii) 5-year tax holiday for new power projects.
(iii) Guaranteed 16 per cent rate of return on paid up and subscribed capital.
(iv) Handing over of distribution networks to private companies.

The Government of India has decided to extend counter-guarantees to the 8 fast track Independent Power Projects (IPP). This counter-guarantee agreement included power purchase agreements signed with the State Electricity Boards and involvement of the Power Finance Corporation. These 8 fast track projects were cleared keeping in view the immediate needs of the power sector in the country. (Subsequently in 1995 this procedure has been replaced by competitive bidding)

Petroleum Sector

In the petroleum sector the major reform that were brought around was corporatisation of Oil and Natural Gas Commission. The newly established company, viz, Oil and Natural Gas Corporation Limited, was asked to raise finance from domestic and foreign capital markets. The programme of disinvestment of its equity by 20 per cent was started. The government is now introducing a system of continuous bidding for exploration contracts, both off-shore and on-shore. Private companies have also been invited for exploratory services to improve the prospectivity of oil reserves. During 1993-94, contracts were also finalised with private companies, both foreign and domestic, for the development of medium and small sized oil fields.

The government has allowed imports and distribution of certain petroleum products like domestic LPG and kerosene by the private sector at market prices to stimulate new investments and promotion of operational efficiency. Private and foreign companies are now allowed to import oil, exploration and production in joint ventures with ONGC and OIL and also in the refining of petroleum products. The domestic market in lubricants has also been opened to foreign collaborations during the year.

Coal Sector

In order to encourage private sector investment in the coal sector, the Coal Mines Act 1973 was amended with effect from June 1993 for operation of captive coal mines by companies engaged in the production of iron and steel, power generation and washing of coal in the private sector. Import of coking coal under Open General Licence is being allowed and the import tariff has been slashed from 85 per cent to 35 per cent. Captive coal mining blocks have been identified for 11 power generating boards/electricity boards and for three companies in the iron and steel sector.

Major Area of Concern—Energy Conservation

The only area where efforts have still to be promoted aggressively and on intense scale is the area of energy conservation. Though there has been an emphasis on the need for energy conservation, but most of the time the thrust has been on energy supply side management rather than the demand side management as that could be planned and administered easily by the government machinery.

ENERGY SITUATION IN SELECTED DEVELOPING COUNTRIES IN THE ASIA PACIFIC REGION

The World Bank has predicted that 5 countries in the Asia Pacific Region, namely, China, India, Indonesia, Korea and Thailand, account for more than 80 per cent of the energy consumption of all Asian developing countries, whose growth rate of energy consumption during the 1980s was higher than any other part of the world. The study projects that the energy demand in 2005 in the 5 major Asian developing countries together will be 2.5 to 3 times as much as in 1990. The energy demand patterns of the 5 major countries differ from energy demand patterns in the industrialised countries. In Asia the pattern includes:

(a) high energy intensity,
(b) high energy consumption in industrial sector,
(c) rapid growth and demand for electricity,
(d) rapid modernisation,
(e) huge losses in electricity, and
(f) low energy prices

The World Bank study also projects that given the rapid growth in energy demand these countries should pay more attention to energy conservation. The weighted average energy intensity to a fiscal unit of output of major industrial products such as iron and steel, cement, fertilizer and pulp and paper is estimated to be much higher in the 5 major Asian developing countries than in the industrialised countries: 55% higher in China, 50% in India, 30% in Indonesia, 20% in Thailand and 10% in Korea. The World Bank has made following projections of energy demand in 5 countries.

TABLE 15.1: Projections of Energy Demand in 5 Countries, 1990-2005

	1980	1990	2000	2005	Growth Rate 1980-90	Growth Rate 1991-2005
	(million tonnes of oil equivalent)				(percent p.a.)	
China	369	612	1198	1587	6.2	6.5
India	72	119	213	284	5.3	5.6
Indonesia	18	30	62	88	3.3	7.2
Korea	34	70	135	193	8.2	6.9
Thailand	8	19	44	66	10.2	8.7

Energy conservation has been suggested in general the ideal way for increasing the energy capabilities for improving the economic condition. Asian economic growth having been linked to the demand for electricity requires that over next 20 years considerable additions in power generation capacities are carried out in Asia. This has drawn interest of major power equipment companies in the world. The whole power sector is now targeting on Asia Pacific market only.

REGIONAL COOPERATION IN ENERGY CONSERVATION AND RENEWABLE ENERGY SOURCES

The Asia Pacific region has a number of centres that are catering to energy related issues. They have either focused on a few activities or else on certain areas of application.

Besides the projected electricity demand in the above mentioned 5 selected countries, most of the countries in this region are richly endowed with agro-residues of one kind or the other. Without realising the role of such agro-residues in meeting the energy needs locally and for applications ideally suited for specific situations, emphasis has been laid on a number of energy systems and devices that are not appropriate. A major reason for renewable energy applications not having caught on in the Asia Pacific region, has been the lack of availability of technology and lack of awareness of the potential. The moment one shifts its emphasis from electrification to energisation, it is very simple to realise the importance of such renewable energy sources as agro residues, solar energy, wind energy and energy from urban, municipal and industrial waste.

The Dutch funded project on Briquetting of Biomass implemented jointly by the University of Twente and Indian Institute of Technology, Delhi has established that available equipment could be suitably modified to make it appropriately operational on different kinds of agro residues. These residues include saw dust, coir pith, rice husk and jute stock. India, Myanmar, Thailand and Malaysia are the countries that have already gone in for briquetting of biomass to a substantially high level. Other countries in this region have an equal stake and if the agro residues are appropriately utilised they can form an important component of energy supply.

Similarly, other renewable sources of energy can significantly help in rural energisation in this part of the world. The prosperity in this region may be improved through centralised high cost electricity supply, purchased on the market terms and used for marketable industrial products. There can be no denying the fact that over the number of years in the 21st century the rural poor in the South and South East Asia Region will live with low amounts of energy, most of them without electricity. The world leaders had expressed concern at the rising numbers of poor, malnutritioned and hungry. There was a special focus on South and South East Asia. Expression of concern creates an awareness of the problem but does not offer any solution. Solutions have to be found locally. And, therefore, collaborative effort of governments, experts and people of this region can only find appropriate solutions.

A possible approach lies in developing a combination of conservation of commercial sources of energy and better harvesting of renewable sources of energy. Conservation without promotion of renewable sources of energy will not make much sense on the overall national scene. Conservation of commercial sources of energy is focusing attention only on two sources: electricity and petroleum; and on two sectors of application: industry and household. There has been negligible attention to the agriculture sector, whereas the need for conservation is the highest in agriculture so that the poor farming community earns maximum returns. Agriculture shall continue to be the main area for employment in the South and South East Asia region. It is here that the renewable sources of energy have a much greater role to play in improving the quality of produce, its preservation, processing and value addition. Future energy policies in the South and South East Asia Region shall have to focus on sources that will create larger employment opportunities in the dispersed rural areas for equitable development and avoidance of rural to urban migration.

This type of development is also suitable for the region as the technologies for renewable energy systems have become commercial for a number of applications and the resources are locally available. Institutional finance required for the centralised large systems is difficult to access; and terms for foreign investments under the garb of bankability have scant concern for the needs of the poor but only the soundness of the investment so that money is safe and secure.

CONCLUSION

Larger and effective cooperation amongst the countries of this region has to be promoted to give a sound and safe energy future to the people of this region. This cooperation has to be at the government level, as well as at the industry level.

INDEX